Contents

Abbreviations

ASV	air to surface vessel	Lt Gen	Lieutenant General
BG	Bombardment Group	MSFU	Merchant Ship Fighter Unit
Cdt	Commandant	NCO	Non-Commissioned Officer
Col	Colonel	O/C	Officer Commanding
ETA	estimated time of arrival	OTC	Officer Training Corps
F/L	Flight Lieutenant	OTU	Operational Training Unit
F/O	Flying Officer	PK	Propaganda-Kompanie
F/S	Flight Sergeant	P/O	Pilot Officer
IO	Intelligence Officer	Sgt	Sergeant
LOP	lookout post	S/L	Squadron Leader
LDF	Local Defence Force	Sqdn	Squadron
Lt	Lieutenant	Wg Cdr	Wing Commander
Lt Cdr	Lieutenant Commander	W/Op	Wireless Operator

Acknowledgements

It would be virtually impossible to record the names of all those who helped my research for this book. Some are named in the credits for photographs, but there is still a vast number of people who helped with information, anecdotes and hospitality over the last few years.

I owe a particular debt of gratitude to PJ Cummins of Waterford city who has been generous to a fault in sharing with me his wide knowledge on aviation in Ireland. In the same context, I am indebted to James Stewart of Omagh and Peter Berry, MRAeS, of Ayr.

My thanks also to the staff of: Military Archives, Dublin; Royal Air Force Museum, London; Imperial War Museum, London; the US Air Force Archives in Alabama, and the Bundesarkiv in Germany.

The book has been enhanced by the skill of Ronnie Hanna, editor at Colourpoint Books. I am glad to record the prodigious work undertaken by my wife, Monique.

Finally, I hope this book suitably marks the help of all the above – any mistakes or omissions in it are mine alone.

Preface

Prior to the World War Two an aeroplane was a relatively rare sight in Irish skies. There were of course air displays in the larger towns and the occasional appearance of private flyers and planes from the tiny Irish Air Corps. A proving flying boat flight to Foynes near Limerick would be received by government dignitaries and hundreds of onlookers. Imagine then the excitement in a rural area when a distressed warplane, often a four-engined bomber or flying boat, suddenly appeared circling to find a landing place. Or at night, the sudden shock of a crash followed by explosions and brilliant fires from an aircraft which had struck a mountain. An air accident in one county was often unknown in neighbouring areas because of very strict censorship of news reporting. So, many of the local dramas recorded here would still, six decades later, be news to many.

No matter from which of the warring nations the hapless planes had come, the reaction of local people was predictable. Often before the police or army authorities could get to the scene, they would endeavour to succour and look after the wounded, refresh the survivors, and show due respect for the dead. The latter would be 'waked' by holding a vigil and reciting prayers – help for the quick and respect for the dead was universal. The loss of fine young men in their prime made a deep impression on the population, most of whom had not experienced the horrors of modern war. The present writer, as a member of the newly-formed Local Defence Force (its British equivalent was the Home Guard) witnessed several of these occurrences at first hand.

Readers who enjoy travel in the comfort of today's aircraft, even in the sometimes cramped seats of 'economy class', cannot conceive of the spartan conditions of World War Two aircraft. There was no noise insulation, often freezing temperatures, appaling weather conditions above which today's jets sail merrily along, rudimentary navigation before airborne radar was developed and, of course, the ever present danger of enemy fire from ground and air. Comfort, except perhaps in large flying boats like the Sunderland, was minimal, and even in this type flights of up to 24 hours had sometimes to be endured.

'The day the plane came down' is still remembered in many parts of the Republic of Ireland today, even after the passage of 60 years. Thanks to various aviation organisations and local communities, memorials to the dead are now commonplace throughout the country. Not only are the casualties commemorated by crosses and

'The Devil Himself', a USAAF B-17 Flying Fortress which spent more than a fortnight grounded on a Sligo beach, displays its running total of nine bombing missions and four German aircraft shot down, while the carrot-chewing Bugs Bunny (the squadron's badge) looks on (see page 126).

US Air Force

plaques, but fortunate survivors also. Some of these have returned over the years to revisit the scenes of their survival and the places where their less fortunate comrades are laid – and to renew old friendships. Through ageing eyes, once-young airmen frequently gaze at mountains, lakes and seashores where their luck had held out. What follows in this book is a general view of a little-known sideline of the air war over Europe and the Atlantic. Its effect was palpable on a nation which, though not officially at war, was nevertheless deeply involved in it.

From 1937 until 1949, encompassing the period with which this book deals, the Republic of Ireland was officially known as Éire, the Gaelic name for Ireland. The population generally resent the continued use of this term or indeed the use of 'Southern Ireland' particularly because Donegal is the most northerly county in the island. For most Irish people the preferred name is 'Ireland'. However, the name Éire is used in this publication as it was the official designation during the period in question.

Not all relevant activities are described in this book for a variety of reasons. Many documents have been destroyed: only five per cent of Luftwaffe archives has survived; several brief 'touch-and-go' incidents were not fully recorded in Ireland – indeed some of the latter were not reported to the authorities until the aircraft involved had 'touched and went'. By its nature, the narrative of this book is episodic. It is driven by the accidental events that would, literally, appear suddenly out of the blue.

Donal MacCarron
Gerrards Cross
May 2003

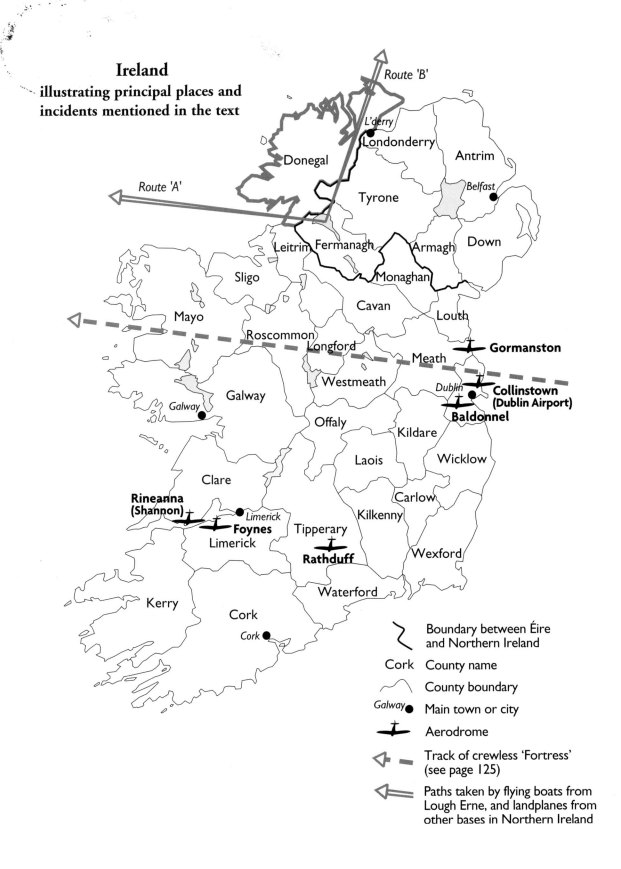

Ireland
illustrating principal places and incidents mentioned in the text

Route 'B'

Route 'A'

L'derry
Londonderry

Donegal

Antrim

Belfast

Tyrone

Down

Leitrim Fermanagh

Armagh

Sligo

Monaghan

Cavan

Louth

Mayo

Roscommon

Longford

Meath

Gormanston

Westmeath

Galway

Dublin

Collinstown
(Dublin Airport)

Galway

Offaly

Baldonnel

Kildare

Wicklow

Laois

Clare

Carlow

Rineanna
(Shannon)

Limerick

Kilkenny

Foynes

Tipperary

Limerick

Rathduff

Wexford

Kerry

Waterford

Cork

Cork

Boundary between Éire
and Northern Ireland

Cork County name

County boundary

Galway● Main town or city

Aerodrome

Track of crewless 'Fortress'
(see page 125)

Paths taken by flying boats from
Lough Erne, and landplanes from
other bases in Northern Ireland

6

1 Guarding Éire

Éire decided to stay aloof (nominally at least) from World War Two. Why? Certainly there was no love for Nazi Germany and the relationship with the country's neighbour, its master for many centuries, could be described as a love/hate one. But pragmatism, rather than emotion, was the keynote.

Before the outbreak of war the magazine *Flight* featured the air arms of all the British dominions. Of Éire it said:

> The use of Ansons in Éire is particularly interesting in view of the British Air Ministry's decision to change the character of the Northern Ireland AAF Squadron at Aldergrove from a bomber to a general reconnaissance squadron charged with the special duty of escorting convoys off the Northern Ireland coast during wartime.

Obviously the editor had no doubt that British and Irish Ansons would be flying side-by-side in the impending conflict. However, this was not to be.

The roots of Éire's neutrality lay deep in history. The original Anglo-Irish Treaty, which in 1921 created the Irish Free State as a self-governing dominion in the British Commonwealth, still provided for the retention by Britain of certain sovereign ports and associated aerodrome facilities. The Free State assured her neighbour in 1936, and again in 1938 when the fortified ports were handed back, that the country would never become a base for an attack on Britain.

To the average Irishman neutrality was perfectly logical and natural, even among committed nationalists who, for historical or political reasons, had always displayed anti-British feelings. The Anglo-Irish were naturally strongly pro-British. However, virtually all elements of the population had a strong dislike of both the Nazi and Communist movements. Also, there were still painful memories of the Great War: although there had been a strong movement against conscription, over 500,000 Irishmen had flocked to the colours and left 50,000 dead on the field. Nobody wanted a repetition.

Almost by definition, neutrality is something negative; however, Éire raised it to the level of a national principle so that its stance was subsumed into the 'sovereignty' which had been hard won just 17 years earlier. The belligerent nations, preoccupied with their own battles, could not reasonably have been expected to spare time or trouble to examine the basic causes of Éire's attitude. They were not above attempting to suborn it when it suited them.

The Irish stance was wholeheartedly backed by all political shades, and there were many valid reasons for this. The partitioning off of the north-east corner of the country was one; another, the very real possibility of internal strife being generated by the still-active IRA if the country supported Britain and, a major factor, the totally inadequate Defence Forces.

In 1939, Irish army numbers were under 20,000, less than half of whom were regulars. In September 1939, Reserve and Volunteer Forces were called up and all were engaged in brisk collective training. Until, that is, the parsimonious Department of Finance advised the government that it could not justify any greater strength –

The Air Defence Command's plotting table was located underground in Dublin Castle before the unit moved to a monastery near Baldonnel Aerodrome.

Irish Independent

Of the 28 anti-aircraft guns available, most were sited to defend Dublin city. This view shows a 3.7in static weapon with camouflage netting.

even though the total force was only at 50% of peacetime establishment. The attitude of those who held the purse strings was that undoubtedly the war would be fought in continental Europe, far from Éire's shores. To save money these financial strategists recommended that some of the reserves be stood down.

When the Allies were routed in Europe in 1940, and the 'Sitzkrieg' turned into the 'Blitzkrieg', there was a clear expectation that the all-conquering Wehrmacht would invade both Britain and Ireland. The laissez-faire attitude changed overnight: reservists were called back and leading politicians of all shades appealed to the nation's manhood to join the army or the newly-announced part-time defence and security forces. The response to this call to arms was astonishing: within 24 hours representatives from every section of the community flocked in their thousands to join up. How to equip them was a major problem: Britain, traditionally the main supplier, was no longer a source.

In the words of the imperialist Rudyard Kipling, "Where there is fighting there's bound to be Irish," and the British forces were strengthened by 180,000 volunteers from Éire who rushed to join one or other of the services. With true 'Irish' logic, however, these still supported the neutrality of their hearth and home. Great Britain's war factories and

aerodrome construction attracted 170,000 workers.

Sir John Maffey, a retired diplomat, was given the carefully-worded title 'British Representative in Éire', and used his skill and long experience to involve the country in the 'struggle of the free world'. Britain did not seem to take account of the fact that most countries, including the USA, remained neutral until attacked. Churchill contended that Éire "is at war, but sulking". Maffey confided to the agitated prime minister:

Hateful as Éire's neutrality is, it has been a neutrality friendly to our cause. I need not give in detail what we have got and are getting in the way of intelligence reports, coded weather reports, prompt reports of submarine movements, free use of Lough Foyle, and to the air over Donegal's shore and territorial waters. What I must stress to you is that I am constantly asked to put proposals to the Éire Government which could not be put in an unfriendly manner. If we could say: 'Éire can go to the devil, we don't want the ports, we don't want anything', my task would be easy! Today, I have on my table a proposal from the Admiralty to install boom defences in the Shannon Estuary, a scheme from the Air Ministry for a corridor of considerable extent over Éire territory and territorial waters enabling the full

Very close Army cooperation: Air Corps Lysanders fly at zero feet over the Curragh in 1940.

By 1942 the Local Defence Force had a more workmanlike green battle-dress, complete with web equipment and bayonets for the US Springfield rifles.
Military Archives

development of Lough Erne as a flying base, plans for the extensive use of Foynes Flying-Boat Base to replace Poole . . .

In Éire the verdict on the ground was that the country would do well to stay out of the war, especially as it was recovering from a long struggle for independence, a civil war with subsequent costly and painful reconstruction, and was dogged by mass emigration. It was against this complex background that the neutral/ un-neutral permanent aircraft carrier in the western sea coped with warplanes from every quarter. The historian Roy Foster came to the following conclusion in 1988: "Neutrality was an affirmative rather than a negative stance. It nourished national pride and independence and, for many citizens, moulded their image of the young state."

Neutrality is, however, something which needs defending but Éire could do little when belligerent aircraft invaded its neutral air space. Though the limited anti-aircraft guns occasionally fired on intruders, the gunners were more concerned with urging them to sheer off – unless bombs were being dropped, in which case closer aim was taken. Otherwise, the policy was not to engage intruders, unlike neutral Switzerland. Swiss fighters were sometimes involved on equal terms in dogfights with the Luftwaffe – occasionally Messerschmitt against Messerschmitt. The Irish Defence Forces had always been at the mercy of the penny-pinching civil service and had no viable air defence. Perhaps this is somewhat harsh on the Department of Finance because the worldwide depression of the 1930s and, in particular, the 'economic war' with Britain had given the army a very low priority in budgetary matters. It was just about sufficient for internal security.

In 1936, the Chief of Staff had recommended a three-year purchasing programme, stressing that the Air Corps and anti-aircraft defence should be given special attention. Nothing happened, and though the Munich crisis of 1938 provided a slight spur, by then it was a case of 'too little, too late'. Military stores began to trickle through in 1939 but the chief supplier, Great Britain, had to look to its own needs. After the fall of France in 1940, invasion loomed from either or both sides – Britain wanted the 'Treaty Ports' returned and Germany would undoubtedly include Ireland in its Operation Sealion plan for the invasion of the UK. So the expanded army found itself keeping watch on the state's entire perimeter, including the Northern Ireland border.

The Air Corps had a mere handful of useful service aircraft, though its trainers were adequate enough to allow 50 pilots to gain their wings throughout the war. But in 1940 the government, now panicking, asked Britain to train 100 fighter pilots and said that it would gladly purchase the necessary warplanes for them. But there was no chance of this happening. Indeed, seven Avro Ansons, ordered pre-war to add to the nine already on strength, though finished in Irish colours and ready for imminent delivery, were retained by the RAF. To boost the First Fighter Squadron cadre, a dozen Gloster Gladiators had also been ordered but they too were not delivered. The US Navy had promised to provide one or maybe two squadrons of Grumman Goose amphibians in 1939 for coastal patrolling, but the Department of Defence quibbled at the very reasonable price quoted and British political intervention in Washington later put paid to this opportunity.

Though guns and aircraft were woefully short, passive defence was highly developed. Air Defence Command (ADC), initially based in two Dublin sites, moved finally to a monastery near Baldonnel where all aircraft movements over the country and its territorial seas were closely monitored. Sightings by coastwatching stations, air patrols, and other security forces were collated by ADC. Annual totals averaged about 18,000. The height and direction of all observed aircraft were broadcast on a very wide frequency in the hope of dissuading the belligerents from using Irish air space. This obviously was more useful to Britain than to her adversary. Every field capable of taking a troop transport or a glider was blocked in June 1940 by the hastily-formed Local Security Groups A and B – the former becoming the armed Local Defence Force (LDF) at the end of the year.

As part of passive defence of neutrality and the safety of the population, the Air Raid Precautions Act 1939 provided for comprehensive schemes in all the main cities and towns – but this did not affect rural areas, with one or two exceptions. In Ballaghadreen, County Mayo, some zealous soul went around putting tin canisters over the bulbs of every street lamp in the town. This resulted in the small town being plunged into darkness except for a small restricted circle of light under each lamp. The probability of any of the warring nations desiring to bombard

Wartime watch on the west: three lieutenants peruse a map during pre-flight briefing – left to right, Billy Ryan, John Walshe and Bill Karney.

Irish Press

Ballaghadreen was very remote indeed.

There was always the danger of aircraft under fire offloading bombs indiscriminately, which was a factor Air Defence Command always considered. Generally, intruding aircraft made off when challenged and though there was no repetition of the major German attack on Dublin, bombs were frequently jettisoned countrywide. The Dublin attack, on the night of 31 May 1941, caused considerable damage and casualties. Midway through the war, the gunners were ordered to refrain from firing on Allied aircraft in daylight if there was no provocation.

Despite minimal resources, the Air Corps was doing its best to guard the country. A couple of days before the outbreak of war, a newly-formed cadre of a Coastal Patrol Squadron began operations from Rineanna in County Clare where the Atlantic Eastern Terminal was being constructed (this, post-war, became Shannon International Airport). The unit had four Avro Ansons, a pair of Walrus amphibians and a runabout Cadet biplane; its manpower totalled 74 in all ranks. The airmen were billeted in huts similar to those occupied by the construction workers who were working merrily away on the concrete runways. For over a year there was no hangar. Aircraft had to be picketed in the open and flown back to Baldonnel, the main base, for

20-hour inspections until a large three-bay hangar was erected to shelter them.

The unit's mission was to patrol the west and south-west coasts, up as far as Donegal and down to Wexford. Early on, the heartily-disliked American Minister in Dublin, David Gray, requested that the coastal patrols be terminated short of the Wexford wetlands or 'slobs' – apparently, the aircraft were disturbing the wild geese feeding there, so spoiling his wildfowling activities. This frivolous complaint got the reply it deserved.

In order to deter marauding and meteorological reconnaissance Condors of the Luftwaffe from taking shortcuts across the south-west, details of their courses were broadcast on a wide wireless band. Acting on this news, RAF aircraft would often be observed in hot pursuit. Occasionally there were close encounters between Irish and German aircraft, but different radio frequencies precluded the transmission of warning messages. The Irish airmen would frequently observe ferry flights on course for Northern Ireland or Scotland.

Patrols from Baldonnel covered the east coast with biplane fighters, and later with Hawker Hurricanes and armed Miles Master trainers, but no shots were fired in anger. One pilot recalls the reaction of an RAF crew, off

Deserted now, a typical concrete lookout post is pictured.

PJ Cummins

course south of the border, when he indicated they should vacate Irish airspace. The British pilot showed that he was anxious to comply by pressing hard on the throttle with his flying boot – the incident closed with grins all round! The Irish pilot also encountered the Luftwaffe over Dublin:

> To reinforce some warning shots from the AA guns, I was sent after a Dornier which simply waggled its wings and zoomed away like a bat out of Hell – or so it seemed to me in my little Gladiator.

Perhaps the strangest intruder encountered by the Irish airmen was an RAF Miles Master piloted by a heavily-moustachioed type accompanied by a pretty blonde WAAF. The two pilots merely exchanged a cheery 'thumbs up' and went about their respective missions. This was typical of the non-belligerent attitude towards harmless-looking strays.

The 'Saygulls': Ireland's 'radar' chain

Conscious of the oncoming conflict, and remembering the U-boat war 25 years earlier, the Defence Forces selected some 83 sites for lookout posts (LOPs). These sites gave a good view over long stretches of coastline and overlooked bays and ports. In wartime the posts would be manned by coastwatchers who dubbed themselves the 'Saygulls' – perched as they were on promontories and headlands. For the most part these were men who made their living from the sea, many of them born beachcombers who regarded the littoral as their preserve.

Bell tents, often blown away by savage winds, were the initial nests for the Saygulls before concrete huts were erected. The coastwatchers were unarmed, but equipped with binoculars and telescopes and, like the air patrols, they acted as front-line sentries who watched to the seaward for invasion forces, shipwrecked mariners, wrecked vessels, and deadly sea mines. They were given minimal military training, and were the butt of many jokes by the fighting troops. But their officers had greater respect for these men:

> The Saygulls were generally loose-limbed gangling lads who would occasionally be seen on a barrack square for what was laughingly called 'retraining' – they not having been properly trained in the first place. But no fully-trained soldier would have got into the places on cliffs or slippery foreshores which our men could handle with ease. They were competent with aircraft recognition and were up to date with the latest types, even before the official silhouettes were issued to them.

Their preserved log books demonstrate their dedication as Ireland's 'radar'.

Every post had a telephone – a not inconsiderable installation in the Ireland of that time and particularly difficult because of the remote location of the LOPs. Eventually, communication had been refined to the point where the LOPs could talk to a special section of G2 (Army Intelligence), superceding a more roundabout chain of command. The brief given the coastwatchers was to monitor every possible aspect of military intelligence. After the fall of France, marine movements played a lesser part in the coastwatchers' duties which now focused on aerial activities. At night-time, specific LOPs and Garda stations had an open line to Air Defence Command which, much to the satisfaction of Britain, did not just deliver 'passive reports' but rather full plots of all aircraft over a period of time. The Saygulls were on 24 hours' duty in shifts and reported regularly to their control centre, including on-the-hour signals of 'zero activity' – a distinct improvement on the early days when they were hurriedly setting up bell tents from which to observe the 2,800 miles of coast. The Saygulls were witness to the increasing activity that would characterise the aerial 'fall out' on all but the Midland counties. The situation of the Saygulls is well described by the Ulster author Breege McCusker, following a postwar visit to an LOP:

> The climb I made to LOP No. 72 on Malin More was both frightening and exhilarating. When I stood in Glencolumcille in late August 1949, I looked out at the miles of Atlantic stretching before me and realised the incredible job these men were asked to do. The wind was howling around me, and I had difficulty in keeping my

At the request of the USAAF, headlands, usually close to lookout posts were identified with 30-foot-high white numbers, illuminated at night and located beside the word 'Eire'. This view is of a navigational signpost on the Wexford coast.

balance. Below was a steep cliff. As I looked at the small post that was the home of James Malloy, Colum Mockler and their comrades, I could well imagine what it was like for them in 1941. It did not look very inviting: the post measured only 9ft × 7ft, built of concrete blocks with a fireplace and a big bay window looking out onto the sea. For fresh water the men had to climb half-way down the cliff to get supplies from a spring. The Saygulls became very proficient at recognising different aircraft from the sound of their engines. From the LOP in the northwest, all patrolling RAF planes were logged, both outwards and homewards, and often the Northern Ireland bases received information from Éire before they knew officially that one of their aircraft was missing.

2 Opening rounds

In the last days of peace, Squadron Leader Michael Collins had been at the Marine Aviation Experimental Establishment at Felixtowe putting Sunderland L2158, the first production model of this flying boat, through its paces: high-speed taxiing, take-offs, checking engines, cameras and the two gun turrets. He returned to Pembroke Dock in Wales and on the fateful 3 September 1939 set out for Stranraer in Scotland in company with two other flying boats, unaware that World War Two, though expected, was imminent.

On that wet and blustery day, a portent of things to come in Éire arrived in the shape of a Saro Lerwick twin-engined flying boat. The Lerwick landed outside Dun Laoghaire, the packet steamer port near Dublin, and stayed briefly. Shortly afterwards L2158, with Collins at the controls, had also been thrown off-course by the foul weather, and alighted north of the city at the resort of Skerries. The pilot and a crew member rowed ashore to establish their whereabouts, unaware that their Prime Minister, Neville Chamberlain, had declared war a few hours earlier.

Collins was invited to lunch at an officers' mess where he verified his flight plan. After due deliberations and a good meal, he was allowed to depart with a parting gift of a case of Guinness – he could hardly go wrong in an army whose first Commander-in-Chief's name he bore! In the meantime, many sightseeing boats had rowed out and circled the Sunderland, their occupants being further enthralled when the Lerwick landed again and towed its bigger brother out into deeper water, from where both took off into a clearer sky. International agreements covered warships which had to enter neutral ports for repairs and these were granted a certain number of hours – so both flying-boat crews were treated as 'distressed mariners'. A formula had still to be found covering all types of belligerent aircraft.

On 12 September an aide memoir was presented to the British Foreign Secretary by the Irish High Commissioner in London prohibiting the use of Irish territorial waters and airspace; a similar message was given to the German envoy.

This prohibition was confirmed during a meeting between the Irish premier Eamon de Valera and Sir John Maffey, the British representative, on 14 September during which a second incident occurred. Another Sunderland, this time from 228 Squadron at Pembroke Dock, alighted on the wide waters of Ventry harbour, County Kerry, and as the two men conversed, a telephone message informed 'Dev' of the episode. He plaintively expressed to Maffey the dilemma which these incursions were causing. Despite the strict press censorship, the whole affair of the earlier flying-boat landing had provoked a great deal of public comment.

Meanwhile, down in Kerry the pilot, acting F/L EJ Brooks, and a fitter rowed ashore in their dinghy and a passing motorist brought them into Dingle, where the cause of the trouble, a broken fuel pipe, was repaired in a garage. De Valera said that this Sunderland crew would have to be interned but Maffey replied that in view of the more lenient treatment in the earlier case, the newly-shipwrecked aircrew should also be allowed to depart. A solution to the argument was found when a second telephone call announced that the aircraft, after eight hours on the water, had done just that. Maffey believed that the timing of the two phone calls had been staged to give maximum dramatic effect. In the wake of these incidents, internment for the crews of seaplanes and landplanes now looked very likely.

The cool captain of the Ventry Sunderland (N9023), who had pressed a car owner and a garage mechanic into service to get his craft away, was still in command when it damaged U-55 ten days later. The submarine had already been hit by RN ships and the Sunderland's contribution caused the German captain to scuttle his vessel. The Sunderland and the Royal Navy shared honours for this kill.

A year later, in August 1940, the practice of internment for downed airmen was put into operation, to the great annoyance of the crew of a Luftwaffe Condor which had a remarkable escape when their aircraft force-landed. A month later an RAF fighter pilot who had shot down a Heinkel was to become the first Allied internee. A pattern

Sunderland L 2158 at Skerries.

Lewick L 7252 at Skerries.

had been set but it was operated differentially as time went by, as described in later chapters.

In those early days, erring landplanes were given lenient treatment too, particularly if able to get away under their own steam. A Hampden, returning from a raid on Germany, was off-course and low on fuel but luckily found the Curragh Camp where, in 1913, the Royal Flying Corps had established an instructors' school. A landing strip had been maintained with aviation fuel on tap. The crew had spotted a searchlight mounted on the huge water tower which dominates the army town. Its operator persisted in following the aircraft, blinding the pilot until the light was

eventually beamed on the landing strip. Down on this safe haven, a young member of the crew excitedly described his exploits over the Reich until his comrades told him to hush-up as they were in neutral Éire. The flak and bullet perforations on the Hampden told their own eloquent version of the story.

Initially the crew were astounded when soldiers wearing German 'coal scuttle' helmets surrounded their aircraft. This headgear was the old-style army issue which Churchill so disliked that he had allowed quantities of the British 'steel toby' to be supplied, unlike other military stores which Éire so badly needed. The Hampden crew were

This is the first production Sunderland which was one of the two flying boats which alighted in Eire waters on the first day of the war – its crew being unaware of this fact. Its final service was as 'KG-M' of 204 Squadron based at Bathurst, West Africa. Its appearance since the landing at Skerries has altered. It now has standard camouflage and its fuselage is topped off with 'stickleback' ASV radar aerials. It was eventually lost at sea on 17 August 1942.

Imperial War Museum

Small boys scramble over Ventura AJ460 which finished its forced-landing run just feet away from a stone wall.
Mrs Ursula Teague

refreshed in one of the officers' messes while their aircraft was topped up. The Hampden then departed under the 'distressed mariners' dispensation which was still being applied in the absence of any other regulation. The Hampden crew was oblivious to a less lenient regime which was about to be implemented: not far away from the airstrip was the site of what would soon become the internment camp for many fliers when firmer rules were introduced a few months later.

The Water Jump

The North Atlantic was a critical link during World War Two between the vast output of American aircraft factories and embattled Britain, and later to the US air forces in Europe. David Beaty, pilot and popular author, had flown across the 'steep Atlantic stream' hundreds of times, both in war and peace. His epithet for this route was the 'Water Jump' and it certainly was an obstacle of major proportions in his time.

The first attempt to ferry aircraft was in November 1940 when, after intensive training, crews from Canadian Pacific airline were given 21 Hudsons fitted with long-range fuel tanks. Crews were a mix of Americans, British, Canadians and one Australian – the redoubtable Donald 'Pathfinder' Bennett who led the first flight of seven from Gander in Newfoundland. To everyone's surprise all seven arrived safely in Northern Ireland to join Coastal Command, as did the second group a fortnight later. One Hudson of the third

group crashed, but the crew escaped unscathed – 20 out of 21 aircraft had leaped the Water Jump, good odds in view of the danger of this obstacle.

The Water Jump, however, proved a leap too far for some aircraft types such as the Lockheed Ventura – the Hudson's bigger brother. In September 1942, a batch of seven Venturas arrived in Maine from the factory in California. Over a period of a week they were fitted with overload tanks, and fuel flow, consumption, and other final tests were carried out before the aircraft were flown by RAF crews to Gander. Despite these modifications, the Ventura was just capable of completing the ten-hour flight to Prestwick in Scotland and then only if wind conditions were near perfect. As the seven crews waited for a favourable met report, they experienced the frustration of watching other aircraft, mostly Hudsons and four-engine types, taking off for the UK after only an hour or two on the ground.

After four days of waiting, conditions seemed right and the seven Venturas trundled out onto the runway. Like its fellows, Ventura AJ460 was laden with three and a half tons of fuel in its standard and additional tanks. It had to accomplish the crossing in ten hours, and by dint of judicious cooking of the Dalton Computer, the crew managed to shave this down to nine hours and 59 minutes. The flight took off on the night of 29 September 1942, night-time giving it freedom from the threat of marauding Luftwaffe long-range fighters and the advantage of a

Lieutenant Jim Teague, an engineer officer, and two NCOs by the door of Ventura AJ460.

Mrs Ursula Teague

daylight landfall at the end. Two of the seven just made it across Donegal Bay and into Northern Ireland, another pair were never seen again. Two more got to Prestwick and one, Ventura AJ460, crash-landed in a field in County Mayo.

The navigator of AJ460 recalls:

> We flew between two layers of cloud, so star shots or drift sights were impossible; the radio packed up within an hour (we really should have returned but we were anxious to get home); even worse, the forecasted winds turned out to be very wrong and we encountered strong headwinds, despite the prognosis of the Gander Met man – who was rated 'a genius'. After ten hours, with the fuel gauges almost at zero, we descended below the 300-feet cloud base and saw a line of cliffs and mountains including, it transpired, Ireland's holy mountain, Croagh Patrick, rising to over 2,000 feet. Most of the surrounding area had tiny fields divided by stone walls, but we had to get down somewhere, anywhere, as the fuel gauges were now showing empty. We came in, with wheels up, onto a large sandy seaside field known locally as a 'duach', but our duach had been obstructed with piles of anti-landing rocks. Despite everything our pilot, Roderick Powell, got us down OK.

Thanking their lucky stars and the skill of their pilot, the crew, unaware of their exact location, now witnessed an unusual vision. Again the navigator picks up the story:

> Approaching us we saw a reincarnation of Don Quixote and Sancho Panza – a huge tall man, in a long black coat and a bowler hat, riding a small donkey, with his feet trailing on the ground. In attendance, a few yards behind, came a small boy. This apparition came up and informed us that we were in Mayo in neutral Ireland, so we did not know what to expect. Dying for a cigarette, but with no matches, we asked him for a light (there was no danger from fuel now) and he said he would send the boy to fetch some. Next to appear were two members of the Garda, who told us to raise our hands and checked us for weapons before formally arresting us. Querulously, we asked what would become of us and were reassured by being told that we would be interned in Dublin Castle but let out every Saturday night to go to a dance.

The Air Corps had carried out some repair work but dissuaded the crew from attempting a take-off, and the aircraft was taken by road to Northern Ireland. Eventually the airmen arrived at Prestwick, their original destination, where they were de-briefed and had to sign a paper swearing not to divulge how they had 'escaped'. Ironically, pilot and co-pilot were told to report to a PR officer who

Landfall Ireland

Following the loss of Sunderland 'P for Pluto' of 201 Squadron in County Clare, 'O for Oceanus' set out to reconnoitre the crash site. Its skipper, S/L Derek Martin, is seen standing on the bow of his aircraft.

Wg Cdr Derek Martin

took photographs of them against a background of an aircraft which appeared in the *Sunday Mirror* with the caption, 'They're in from the US'. A supporting story told how they had gallantly delivered an aircraft to help the war effort. Of course they did try hard, but the actual Ventura was still in an Irish field.

The pilot of the Ventura which had got as far as Limavady in Northern Ireland was asked on arrival if he was a new student for the junior course being conducted by the OTU there. He replied haughtily that he was not and had, in fact, just flown the Atlantic. Despite this experience, a few weeks later, he and other survivors of the flight found themselves back at Limavady to complete their training, despite proving that fledglings with lots of luck and guts could already master the Atlantic!

AJ460 was fully repaired and eventually returned to service with Nos 60 and 13 OTUs, before finally being struck off charge in July 1945. Lockheed Ventura AJ460 from 21 Ferry Control had realised its destiny.

Most of the ten other Allied warplanes which came down in County Mayo were not quite as lucky.

Sunderlands

The Sunderland which appeared at the start of this chapter was RAF Coastal Command's staple maritime patrol aircraft throughout World War Two. It was a large, well-armed flying boat capable of 18-hour-long patrols. It was derived from the C class 'Empire' flying boats of the

mid-1930s and was a crucial factor in the Battle of the Atlantic which threatened to starve the UK into submission in 1941–42.

On a miserable day in December 1941 Sunderland 'P for Pluto' of 201 Squadron based on Lough Erne had been out on anti-submarine patrol since 4.00 am. Its redoubtable captain, F/L Grant Fleming DFC, describes how the patrol ended:

> In the evening we were running short of fuel and could not make it back to base so I decided to try and come down just off the coast. I succeeded in doing this, but lost the float on the port wing and I found it necessary to take off again. I was then undecided what course to adopt, but finally came down on the sea some miles further south, but as I did this the outer port engine broke off and caused the aircraft to list and sink. I ordered the crew to abandon the plane – I was the last to leave but was carried under water and had to dive deeper to get clear of the wreck. When I came to the surface I managed to climb on one of our rubber dinghies but was almost immediately washed off and had to keep myself afloat the best way I could. After what must have been a

Six of the seven Sunderlands of No 201 Squadron overfly its Castle Archdale base on 1 December, 1941. The leading aircraft is "ZM-P" (W3988) flown by F/ L Fleming, DFC who was to survive the aircraft's crash into the sea off Co Clare two days later. The left-hand aircraft at the rear is "ZM-Q" (W3977) which was lost in the sea off Donegal on 5 February, 1942. Only two of the aircraft seen here survived the war.

James Stewart

couple of hours I felt sand under my feet; I had been washed ashore by the breakers. While I was in the water I saw Sgt Masterson trying to help Sgt Bennett, but Bennett became panicky and Masterson had to let him go. Two people took me to a house nearby.

The Sunderland had crashed at 6.30 pm off the west coast of Clare. From his bed in Mallow Hospital, Fleming was asking about his greatcoat and cap which he hoped might have turned up when the wreck floated ashore, but neither was found. His assertion that two of the locals who had assisted him out of the water had pinched his watch was completely untrue, though he insisted that this supposed act of banditry had started his anti-Irish animosity which he constantly demonstrated in the internment camp to where he and the other survivor were bound. The endeavours of his rescuers were recognised by awards for bravery.

Salvage work was inspected by a British representative who showed particular interest in the retrieved bombsight which he asked to have destroyed. To the Air Corps there was nothing special about this item because it was a Mk IX, only slightly different to the Mk VIII which the Corps itself used. There were two further obviously secret items buried in the sand underneath the wreck and these also were demolished on request, but not before an Air Corps engineer had made sketches of them.

Between the tides, the rear part of the aircraft was excavated by the LDF, the Gardai, and the 13th Cyclist Squadron, all on the lookout for ammunition. Later a salvage team arrived and one of its members described the scene:

> When we arrived in the nearby town of Kilrush we were accommodated in an old convent. As there was no road leading to the crash area we had to make our way through sand dunes and when a storm blew up we had only the roar of the sea to guide us to the site. The Sunderland had apparently lost both wings and overturned in deep water so access to it was restricted to the time between tides. As the aircraft was on its back we had to go down a stairway to get to the upper deck and the flight deck!
>
> We had been working for some time when a high-ranking Army officer and a member of the British Embassy in Dublin arrived. They took a look inside and when they reappeared they asked for a pickaxe and went back into the aircraft from where we heard loud noises and banging. Finally they resurfaced and returned the pickaxe and thanked us. We later discovered that a secret bomb-sight had been fitted to that particular aircraft and the British authorities were making sure that it did not get into the wrong hands. We continued our salvage for a couple of days. However the bulk of the aircraft was left behind in the sand.

Another 201 Squadron Sunderland (ML743) crashed into the hills near Killybegs in March 1945. The aircraft was completely destroyed by fire and there were no survivors from the crew of 12. It was plainly off course, but again the actual cause of the disaster was never established. The captain was experienced and the navigator was careful and competent. At the time the Atlantic weather was treacherous and many aircraft had to be recalled to base on this account.

On 25 May 1943, Sunderland DD846 from RCAF 422 Squadron was cruising over the archipelago of islands off the Mayo coast when it apparently touched the top of Croaghmore which, at more than 1,500 feet, dominated Clare Island. The flying boat plunged into the sea and all crew were lost. There was some evidence that the aircraft was in trouble and that the brief contact with the mountain was not the major cause of the crash. When the bodies of the crew were recovered it was found that they were wearing their life jackets and had discarded their boots as if expecting a ditching.

This accident was observed by the lighthouse keepers on the island and they, with others, attempted a rescue using the local type of boat, the curragh. This light canvas-clad type of fishing craft is still to be seen today skimming over the waves. When the boats got to the airmen all were past assistance. The Sunderland had been returning from an anti-submarine patrol to its base at Lough Erne, manned by a crew of four RCAF men and six from the RAF; seven bodies were recovered. The family of one of the Canadian pilots, F/O EF Paige DFC, erected a plaque on the island in 1996. Unfortunately, this was not the only Sunderland from the Canadian squadron to be lost in the misty weather conditions off Ireland.

Not all the Sunderlands which came to grief on or off Éire were based in Northern Ireland. Number DW110 of 228 Squadron, which had set out from Pembroke Dock in Wales at the end of January 1944 on a U-boat patrol off the French coast, ended up – high up – on Donegal's Blue Stack Mountains, with disastrous results, only three of its 12-man crew surviving. Sergeant James Gilchrist (who was later commissioned) tells how weather was the culprit virtually from the start:

We had been flying through the most appalling rainstorms; none of the crew until that time had ever been air-sick, but we all admitted now to feeling very queasy. We completed our 12-hour patrol and were on our way home when we were instructed to divert to Lough Erne.

The bad weather continued as we left our patrol and

An evocation by Donegal artist Johnny Boyle of the last moments of Sunderland serial DW110 from Pembroke Dock, which crashed in the Blue Stack Mountains in Donegal.

J Boyle

set course for the west coast of Ireland. It had been the intention to cross the coast of Ireland at the south end of Donegal Bay giving us the shortest route before tracking direct to Lough Erne. However, we had been unable to get a reliable fix for some time. We had been continually in cloud bucketing around in the rain. The visibility was almost zero and it was very dark. I was in the rear gun turret, cold and tired and looking forward to the end of what had been an almost 13-hour roller coaster ride. As it turned out we finally crossed the coast a few miles north of the intended track, unable to pinpoint our exact position. We were all looking for recognition points. Over the intercom I heard Freddie Copp, who was in the front gun turret, say, "I think I can see high ground ahead – I'm sure it's high ground." That was the last thing I heard.

The Sunderland had flown into the rain-soaked Blue Stack Mountains in the south of the county and it would appear that Gilchrist and a mid-upper gunner were fortunate in being faced backwards, though they suffered considerable injuries. The survivors sheltered under a large rock and waited from midnight to daylight, when they located another severely injured casualty. Gilchrist and the gunner made their way painfully and slowly down the mountain for six hours before they found a remote cottage

The last patrol: a Sunderland from 201 Squadron RAF flies out from the Donegal Corridor at war's end, June 1945.

201 Squadron RAF

where a woman and her children tended to them. There was no question of internment for the survivors and an RAF ambulance with doctors and crew in civilian clothes arrived and transported the injured across the border and into hospital.

Forty-four years later, having retired with the rank of squadron leader a few years before, Jim Gilchrist and his family were invited to return to the crash site where he unveiled a memorial plaque for which he had chosen a line from 'The Fallen'.

At the end of the war many of the Sunderlands based on Lough Erne were sold as scrap to Belfast firms; others were towed to the centre of the lake and their seacocks opened while marine craft men used fire axes to speed them to the bottom.

A faller on the Racecourse

In 1910 the newly-established Irish Aero Club held a highly-successful display at Leopardstown Racecourse. The aviators who flew Bleriots and Farmans would have been amazed to see a 1941 visitor to their flying field. It was a Bristol Beaufighter of 252 Squadron RAF, the first unit to be equipped with the type. The squadron had been sent to Malta at the end of April 1941 to give long-range fighter escort to a very important convoy, coded Operation Tiger, delivering 250 tanks through the Mediterranean to Egypt. General Wavell was desperately asking Churchill to rush

more armour to his aid: Rommel's men had driven his army back almost to the Egyptian border. A delivery via the Mediterranean, though eminently risky, would save 40 days by avoiding a Cape of Good Hope voyage.

Hugh Verity was one of the pilots who was very busy covering the convoy for four weeks. When he wasn't flying or sleeping he was at the receiving end of frequent bombing of Luqa airfield. By the end of Operation Tiger there were only four aircraft just about fit to fly back to the UK for major repairs. All had an unserviceable radio and minor bomb damage. Hugh Verity remembers:

After stopping at Gibraltar for a night we set off on 22 May for St Eval in Cornwall with an over-optimistic weather forecast. I believe that one of the four made it; one crew ditched and was rescued; one was lost and we ran out of fuel near Dublin. I had asked Sgt Barnett to navigate me to a point where I could let down to see land through solid cloud 50 miles south-west of Land's End. When I saw trees and hedges going past in the cloud at an indicated height of 1,500 feet I knew that something was wrong . . . so I climbed up and pressed on until I could let down through gaps and saw what was obviously

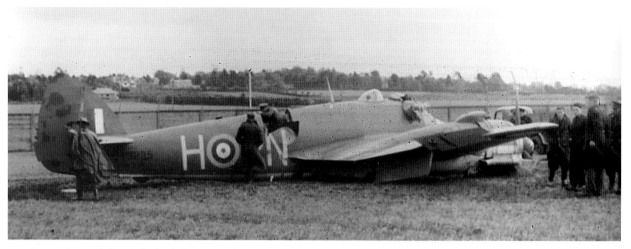

Hugh Verity's Beaufighter belly-landed at Leopardstown Racecourse. Having escaped from internment, Verity was later posted to the special squadron which landed agents in occupied France.

Group Captain Hugh Verity

Back in service, having escaped from the Curragh is Hugh Verity, second from the left, now a Group Captain. He is standing in front of his Lysander which shows his mascot 'Jiminy Cricket' in which he performed many hazardous flights into occupied France, dropping and picking up secret agents.

Group Captain Hugh Verity

a race course. Low on fuel, I decided where better place for a landing, but on my approach I saw that the open spaces were impeded by tripods of poles. I had no option but to put down in an adjacent field, clipping a concrete wall in the process. We climbed out and asked a pretty girl who happened to be there what country we were in? "Sure, you're in Ireland" was the response, on which I climbed back in and destroyed the IFF [identification radar responder].

The crew of three were promptly arrested, deprived of their side arms, and checked out at a first-aid post, being examined by no fewer than three doctors. The only injury was to Verity's forehead which had been grazed by his gun-sight. When the crew arrived at the Curragh, the senior officer there provided drinks and said to Verity that, while he appreciated that the pilots' priority was to escape, *his* duty was to keep him in custody and he trusted that they both understood each other. But later Hugh Verity was the moving spirit behind a successful mass escape, for which he was disguised as an unlikely lady.

Verity went on to command the Lysander Flight of 161 Squadron which was engaged in the hazardous night-time 'drops and pickups' of agents in occupied Europe. All is detailed in his highly successful book *We Landed By Moonlight*. He was awarded the DFC and DSO and subsequently attained the rank of group captain. He passed away in the fullness of years in 2002.

Catalina Catalogue

The Consolidated PBY Catalina amphibian was a US twin-engined aircraft which gave sterling service to RAF Coastal Command in the battle against the U-boats. After Pearl Harbor it served the US Navy with equal distinction. Pre-war, and prior to a later large order of the 'water only' version, the Royal Air Force had purchased a single example, Serial B9630, which set out from Buffalo, in New York state, on 11 July 1939 for its long ferry flight to the Marine Aviation Experimental Establishment at Felixtowe. The Catalina overflew Foynes flying-boat base on its 2,450-mile trip. Coincidentally, a civil example owned by American Export Airlines, coded NC1R997 and named *Transatlantic*, arrived, unannounced, after a 20-hour proving flight from Botwood – its owners were giving notice of their interest in Foynes, much to the annoyance of BOAC and Pan Am, who considered the North Atlantic to be *their* sole preserve!

In the war zone, a Catalina from 240 Squadron got lost and ploughed into Aunagh Hill in County Donegal. It impacted 200 feet below the 1,700-foot summit, its depth charges rolling down to the foot of the mountain. Apparently the crew had lost their way in fog while returning to their Castle Archdale base. An armed guard was put in place to deter souvenir hunters, but not before one old chap had taken away one of the depth charges which he used as a seat for years until the police heard of it and the ordnance men destroyed it. Another local also

A year on from its landing on the Shannon mudflats, this Catalina, now serving with 212 Squadron, is seen on patrol off the coast of Portugal.

RAF

removed a wing and incorporated it into the wall of an outhouse, which is still called the 'Plane Shed' to this day!

Later, on 19 November 1942, another Catalina from Castle Archdale (a Mk IB coded FP202), on charge to 302 Ferry Training Unit, had set out over the Atlantic to search for survivors of a vessel sunk about 500 miles out, but without success. As it was turning for home, radio trouble and engine faults brought it over Ireland where the crew tried to pick up some local landmark. One airman was sure they were over Donegal but, after circling for some hours in the dark, they landed on the Shannon, close to the confluence of its tributary, the River Fergus, bumping slightly on the mudflats but sustaining no damage. The crew, relieved to be on the water after 20 hours in the air, anchored in the main stream. Air Traffic Control at Foynes sent out a launch which located the Catalina whose crew were anxious to remain at anchor. However, they were ordered to follow the launch to Foynes where they were given a hearty army breakfast.

This Catalina was a brand new aircraft, recently arrived from the United States together with eight others. It was armed with four Browning machine-guns, but its bomb load had been dumped at sea. It was allowed to leave after refuelling but the crew now showed great reluctance to depart – perhaps the army's hospitality had something to do with this. One of the more reluctant was jokingly

threatened with internment and said he was quite prepared to accept this, but his comrades hustled him onto the plane! The aircraft was subsequently operated by 212 Squadron and survived until the autumn of 1944, when it was struck off charge.

Later, another Catalina, a Mk IVB coded JX330, made a brief call to Foynes during a delivery flight from Canada, where it had been built by Boeing at Vancouver as one of a run of 200 Catalinas licence-built there. The flying boat had departed on 9 March 1944 using the southerly route from Bermuda as the water bases in the Canadian Maritimes were still ice-bound. On the following day, after 26 hours in the air, short on fuel and with a fatigued crew, it alighted. The army checked that it was carrying no bombs or ammunition and that its two wing-mounted machine-guns were suitably sealed. Another distressed aircraft (coded JX422) from the same production batch would alight at Foynes some months later.

Yet another Catalina, again a Mk IVB, had made two unsuccessful attempts to land in Northern Ireland but its pilot gave up and headed for Foynes. Something of a contretemps occurred when he alighted there. The senior officer at Foynes delayed permission for the crew to come ashore, though they had been in the air for 18 hours and were getting seasick on the rough water. After almost an hour the martinet relented but insisted that one crew

Catalina JX330, whose 26-hour ferry flight ended at Foynes, flew on to its Castle Archdale base. Seen here is its crew returning to shore after an Atlantic patrol.
RAF

member stay on board with one of his men, but soon this jaded crewman too was allowed onto terra firma. Next morning the Catalina took off bound for No 131 Operational Training Unit on Lough Erne where it soldiered on till the unit was disbanded at war's end.

Throughout the war Foynes became a vital link for the Allied cause, despite Ireland's declared neutrality. Military and political VIPs and warlike stores passed through regularly. Security was stringent and the German legation was kept in the dark as far as possible as Allied traffic increased. The Irish authorities never capitalised on this benevolence in terms of political gain, nor in terms of practical matters, such as food and badly-needed military supplies. In fact, apart from landing fees, the country never profited from Shannon's key strategic position during the war.

A few nights before Christmas 1944, the residents of Castle Gregory in Kerry were amazed by a glow of great intensity in the sky. As one of them said, "The light was so great, you could pick up a pin in my front yard! I had heard an aeroplane with rough-sounding engines which stopped suddenly – then there was this great light." This phenomenon proved to be the sad pre-Christmas demise of the nine-man crew of Catalina JX208 who had taken off from Castle Archdale on a 24-hour patrol of the Bay of Biscay. Apparently one engine had gone on fire, which

heavy rain failed to extinguish, and perhaps it was this distraction which caused the Catalina to crash into the mountains.

The ferry flight of another Catalina brought an exhausted crew with empty fuel tanks down on to Lough Gill in County Sligo. Seeing its extended floats as it came in, a local declared that he had seen a man balancing on each wingtip! Lough Gill was WB Yeats' 'Lake Isle of Innisfree' – of which the poet had written, "I shall have some peace there, for peace comes dropping slow". Nearby Sligo was far from peaceful when the crew of five were being entertained right royally in Sligo as the guests of the officer commanding the local coastwatching area – that is until an RAF officer arrived to bring them back into the war! The aircraft (SB273) had come over via Bermuda, heading for Largs in Scotland. After the unscheduled Sligo stop, it went on to Beaumaris for conversion to RAF standards by Saunders-Roe. It served with three units before being struck off charge in August 1945.

In addition to Foynes, two areas of Cork's great harbour had been designated as safe havens for waterborne aircraft, as was Roaring Water Bay on the west coast. Mooring facilities had been put in place at these three locations – but Foynes was the preferred haven for flying boats which had lost their way.

The Catalina which came to rest on tranquil Lough Gill in County Sligo – the poet WB Yeats described it as the 'Lake Island of Innisfree'.

Under the steeple of Cobh Cathedral an RAF Catalina lies at one of the mooring points set down in Cork's great harbour.

Michael Keating

3 Picking up the pieces

The steady stream of warplanes that crash-landed in Éire kept the small Irish Air Corps busy. Apart from keeping its own aircraft flying, despite the shortage of spares, its technicians attended almost 200 sites where incidents occurred, often in the most inaccessible parts of the country. Two, and sometimes three, salvage teams were simultaneously involved working in mountains, bogs, lakes and seashore to reclaim the remains of the various wrecks. Some aircraft were only slightly damaged, but even in the case of major crashes, engines, armaments, and equipment represented valuable returns to their owners. The disposal of ammunition, aerial torpedoes and bombs involved the Ordnance Corps. Apart from public safety, such items could not be allowed to fall into the hands of 'dissident elements'. Also, complete removal of the wrecks would save unnecessary reporting by overflying aircraft.

An aeronautical engineer who was frequently detailed to head up salvage teams recollected:

There was an amusing side to making up a task force. Usually for a major job, the officer-in-charge would require approximately four engine fitters, four airframe men, an electrician, a welder, an artificer and a few helpers. If, as in my case, he was O/C Engine Shops, then I took the best engine fitters and NCOs from my section. Then I would approach the O/C Airframes and the O/C Ancillary Section to get men from them. They saw no reason to relinquish their best men for such 'gad-about expeditions' – so they would, of course, pawn off their less able types on me. Of course, it was an entirely different matter when they drew the job for I would pay them back in their own coin! The same situation prevailed when selecting drivers and cranemen: due to the huge expansion of

When downed Allied airmen were accosted by Irish troops still wearing the old German-style steel helmet, consternation reigned!

Irish Press

26

Bottles of Guinness make a welcome break for a salvage team which has managed to erect a corrugated shelter from the severe weather conditions which often interrupted their endeavours.

Brendan O'Byrne

the Defence Forces, some of the non-technical men could be pretty raw; some were wizards but others could get bogged down on a concrete runway!

I once got my come-uppance, due to overdoing the selection process. My contribution to complete a party for a pilot whom I didn't much care for (air officers were occasionally in charge of salvage teams) represented a dreadful collection: the pilot went to our commander and swore that with this bunch he would not get finished in two months. The boss sent for me and told me to take over the party and get back inside a fortnight. We did the job with two days to spare and I got great work out of some of the 'unchosen'. From the men's point of view, most of them welcomed a trip 'down the country' and away from the army routine of guard duty and other military chores which even highly-trained technicians were subjected to – and which were always a drag on our pure aviation capacity.

Where possible, Allied aircraft were flown back; otherwise they were returned on low-loaders. The low-loaders, called 'Queen Marys', formed part of a consignment of salvage equipment supplied by the British in 1942 together with five-ton electric cranes, jacks and trestles. Properly equipped at last, the salvage teams reckoned that they could move anything out of anywhere.

When the US joined the fray the above engineer officer and his crew were once again 'in the field' or rather, in this instance, in the bog. A Liberator which was en route from Gander to Northern Ireland came up against heavy squalls after 14 hours in the air, and the situation was not helped by the fact that all the navigation aids had gone awry and fuel was getting dangerously low. From his position in the nose, the bombardier kept his eyes peeled for a break in the murk and soon spotted what he considered to be an ideal green field in which cattle were grazing. His pilot scared the beasts away by several low passes and came in with wheels down. But like Alcock and Brown at the end of their 1919 Atlantic flight, the surface proved to be a soft cushion of peat! The undercarriage was swept off and the bomb doors burst open, scooping in tons of turf – an emergency

Above: The 'graveyard' at Baldonnel soon expanded as more and more wrecked aircraft arrived there.

Captain A Quigley

Below: When these two aeronautical engineers, Lt Andy O'Shea and Lt Jim Teague, joined the Air Corps in 1940 they were soon increasingly involved in the salvage of downed aircraft. They are standing in front of one of the Corps' Ansons.

braking system which the designers at Consolidated would never have dreamt of! First up to the aircraft, whose crew had only suffered a variety of scratches, was a Garda stalwart who reassured them that within 72 hours they would be north of the border, and they were . . . after a bout of hospitality during which they got no sleep! But, after their departure, someone had to set about the disposal of their wrecked aircraft. The ubiquitous engineer officer takes up the tale:

The USAAF only wanted the engines, guns, instruments and all removable gear, so that left us with the large air frame which had taken the main impact. There were many problems and hazards, not least from leaking fuel. Though almost 300 gallons were siphoned off, a large quantity had soaked into the bogland, hence, when the uncomplaining oxya-cetylene cutter was at work we had to spray water around him to keep him less flammable. The water came from the numerous bogholes via a pump and battery salvaged from the wreck. My team worked like Trojans, up to ten hours a day including weekends and holidays, and eschewed the help of a nearby garrison when it came to heavy manhandling. We removed the outer wings, engines, instruments, etc and dispatched this valuable cargo to the border.

The fuselage remained to be blown up and I got many quiet approaches

This series of photographs shows the salvage work on a USAAF Liberator.

Above: A USAAF Liberator at the start of the salvage operations.
Mrs Connie Maycock

Below and below right: Two further views of the downed Liberator.

Brendan O'Byrne

Above: The tail guns of the Liberator which came to rest in a mountain bog in Mayo.

Brendan O'Byrne

from local scrap merchants for exclusive salvage rights to what remained. However, I posted notices saying that the carcass, which we had cut into four sections, would be blown up at 3.00 pm on the day when our work was complete. As this hour approached, about six lorries and a dozen farm-carts lined up behind our cordon. When the demolition charge exploded, off went everybody in a mad dash; we even had to fight hard to get our own equipment clear. A pair of local electricians were soon madly stripping switches and wires out of a section of the fuselage (such material was virtually unobtainable) when six burly fellows employed by a rival scrapman picked up the section with the two inside, tossed it onto their lorry, and drove off. Some little way up the road, the scrap merchants kicked out the electricians and proceeded on their way. There were fist-fights and threats all over the place, but eventually everyone simmered down – except the Gardai. During the previous fortnight they had been

running in locals for taking even small souvenirs from the plane – now people were making off with lorry loads of the stuff!

Incidentally, because there was no army barracks locally, I was put up in a luxurious hotel where there was fresh salmon for every other meal: the lads got good digs in Ballina and had a great time with the Mayo lassies! Some jobs had their compensations, and some were easier than others.

The records show that Air Corps teams were involved in 163 major incidents and many minor ones as well.

Whenever it was possible to anticipate or evade the authorities, local people would help themselves to aviation fuel from the wrecks. This, when mixed with paraffin, was a substitute for the real thing when petrol rationing was imposed. Diesel from German aircraft could of course be

Typical salvage engineers at work on the Liberator.

Brendan O'Byrne

An indication of the amount of Air Corps transport required at these crashes is seen here as a crane loads a truck.

Brendan O'Byrne

Irish Air Corps technicians attend to a comrade injured during dismantling operations on a Liberator aircraft down in County Mayo.

Brendan O'Byrne

The 'windfall' Hudson is the background to this Air Corps team which repaired it sufficiently for it to be flown out by an Aer Lingus pilot familiar with the Lockheed type.

Eddy Farrell

used to supplement supplies for farm equipment. Another sideline much in evidence was the manufacture of rings from aluminium and perspex which were used as tokens of affection! As mentioned earlier, whole sections of aircraft were fetched away, either officially or unofficially, to roof farm buildings or be turned into smoke grenades for the army. Another tangible advantage for the Air Corps was various hard-to-come-by spares, particularly from RAF engines.

Hudsons – and a 'windfall'

The Lockheed Hudson was a first-class coastal patrol aircraft which gradually supplanted the Avro Anson in

British service. Seven extra Ansons on order were denied to the Air Corps on the eve of delivery, but the outbreak of war also curtailed Aer Lingus' activities and the company sold its two Lockheed 14s – the type on which the Hudson was based. The company did not consult the aircraft-starved Corps about the sale, much to the annoyance of the latter, but eventually the Corps obtained an example by default.

In September 1941 a Hudson, serial AE577, low on fuel after a transatlantic journey, came into Baldonnel in the morning, breaking its ferry flight to Prestwick. The newcomer was much admired by the local pilots; one who

This Hudson flew in Air Corps' colours throughout the war, then in Aer Lingus' livery, and eventually in that of a Belgian airline.

Peter Walsh

was used to flying the west coast patrol in Ansons described the Hudson as being:

. . . most generously equipped for bombing, fixed and free gunnery, photography, celestial navigation and blind-flying with an auto pilot, a Boulton Paul power-operated turret, de-icing equipment, carburettor heaters, and many more new-fangled ideas. A further benefit was its long-range fuel tanks.

Though the visitors, two Canadians and an RAF wireless operator, refused a period of rest and relaxation, it was also clear that the authorities were slightly anxious to see the refuelled aircraft depart as soon as was practical. This proved to be the visitors' undoing: as it flew northwards it impacted in a 'stuffed cloud' and all were killed on the mountains near Dundalk. Both Irish and RAF investigators reckoned that the pilots had fallen asleep at the controls. The officer who had earlier admired the brand new aircraft was appalled by the tragedy and when he saw the charred wreckage arriving back at Baldonnel he thought that not only would lives have been saved had it been impounded, but a badly-needed aircraft would have been acquired.

Earlier that year, however, a Hudson had been added to the inventory: this was one of the machines from Coastal Command's 233 Squadron based at Aldergrove. While on an Atlantic patrol, it had a brief encounter with a Luftwaffe B+V138 flying boat. Little damage was caused but as it returned in darkness to its base it was seen to 'stooge about', having obviously lost its bearings, as was later confirmed by the pilot who thought he was either over Iceland or

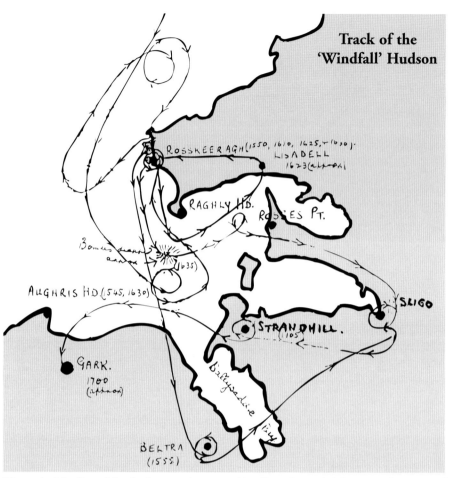

The work of the 'Saygulls' and other agencies was nothing if not thorough. This map, whose original was drawn in coloured inks, shows the convoluted tack of the 'windfall' Hudson prior to its forced landing. Despite darkness, the ground observers noted every change in the aircraft's course.

Scotland. Prudently dropping the bomb load into the sea, he made a wheels-up landing near Skreen, County Sligo. The aircraft pulled up on soft ground after a mere 20 yards. Very little damage was caused: two of the blades on the three-bladed airscrews were bent when the Hudson finished up with its undercarriage halfway down and its tail-wheel well up.

The crew, three pilot officers and one sergeant, had difficulty in leaving the aircraft because one of the civilian rescuers accidentally inflated the rubber dinghy and thereby blocked their exit! They were eventually rounded up by the Gardai and LDF and a military guard arrived, commanded by a young engineer officer to whom the RAF officers took an immediate dislike. They scorned his Germanic-style

uniform which was of the Volunteer Force OTC. For his part he found them "both boorish and unsociable" while their NCO was well behaved.

Soon an Air Corps salvage team arrived and dug trenches into which the undercarriage was lowered, and carried out some minor repair work. The chief pilot of Aer Lingus, familiar with his company's Lockheed 14s, flew the plane down to Baldonnel where it became No 91 in the Air Corps' inventory, having been purchased with spares from the Air Ministry. If the two airline Lockheeds had not been sold off, this pair, suitably armed, together with the 'windfall' Hudson, would have formed a new flight for the Coastal Patrol Squadron. The Hudson did serve briefly in this role but was mainly used as a VIP transport. As for its original aircrew, they were duly dispatched to the Curragh, together with their carrier pigeons from which two messages were retrieved. The birds joined their feathered friends in the army pigeon loft while the crew joined their comrades in the same area. The pilots, who had been so upset by the Volunteer uniform, were further peeved when 'their Hudson' occasionally overflew the internment camp in its new colours.

At war's end No 91, minus its dorsal turret, was sold on, as EI-ACB, for training and air-sea rescue duties to Aer Lingus, which was then expanding. Some minor combat damage was discovered in the rear fuselage, probably caused by the encounter with the Luftwaffe and the airline disposed of it to Belgium. The sole relic of its military service was its dorsal turret which was displayed on the Corps' stand at a Grand Tattoo and Exhibition which marked the end of the war, or as the period was known in Éire, the 'Emergency'.

Colonel Jack Connole as a second lieutenant in the Volunteer Force uniform which upset the RAF!
Mrs M Metcalfe

The 'windfall' Hudson was mainly used as a VIP transport. It is about to be used to fly the Irish Prime Minister, Eamon de Valera (fourth from left), and one of his ministers, Frank Aitken (fourth from right), from Baldonnel to Rineanna on an inspection visit.

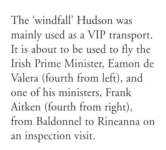

4 Down in Donegal

In pre-war Northern Ireland there were few aerodromes, the principal one being RAF Aldergrove near Belfast. By 1940 the British Airfields Board was urgently searching for level, well-drained sites large enough to take four runways and approaches clear of major obstacles. However, a thorough survey was far from reassuring due to the nature of the terrain:

> In County Armagh it has been found impossible to secure any suitable site ... Elsewhere the dampness of the climate, the wide stretches of peaty and boggy soil, particularly in the low-lying areas, the mixture of undulations and hilly country make even semi-suitable sites rare indeed.

But needs must when the devil drives, and in this situation the devil was represented by the U-boats and Condors which were causing havoc in the Battle of the Atlantic.

Despite the difficulties in 1941, the construction of aerodromes took precedence over all aspects of Northern Ireland's war effort. RAF Coastal Command also required water bases for its flying boats and it was thought that the picturesque Lower Lough Erne would be ideal. Local people had been unaware of the mission of an obsolescent biplane flying boat, a Stranraer from 119 Squadron, over

Lower Lough Erne on Christmas Day 1940. The aircraft was on a 'photo-recce' of the lake's potential, which could add on 100 extra miles of flying boat cover over the North Atlantic. The resulting report was not particularly favourable, but as losses at sea were escalating it was decided to proceed with bases on the shore of the lough. Winston Churchill would have much preferred Lough Erne as a civil Atlantic terminal too, rather than Foynes in despised neutral Éire.

Two main facilities on the lough were built at Castle Archdale and Killadeas by the US Navy, even though the United States was still neutral. Northern Ireland's handful of aerodromes would eventually rise to a total of 19 flying-boat stations, aerodromes and satellite strips. The traffic these bases generated and the numerous transatlantic ferry flights bound for both these and Scotland would cause almost 40 aircraft to come down in County Donegal's rugged terrain and its seaboard. In one single week there was a total of seven forced landings and crashes. The table opposite contains a list of these incidents, giving basic details.

With the airborne arrival of various nationalities from all parts of the globe County Donegal was living up to its ancient Gaelic name, Dun na nGall – 'the Fort of the Foreigners'.

This Coastal Command Liberator (FK222) of 86 Squadron RAF, based at Aldergrove, was forced to land on the strand at Fort Lenan (Inishowen), County Donegal on 18 March 1943. No one was injured and the aircraft itself was dismantled and returned over the border.

Captain Patrick O'Shea

Another view of the Coastal Command Liberator on the strand at Fort Lenan, County Donegal. Salvaged equipment is seen in the foreground.

Captain Patrick O'Shea

Landfalls in Co Donegal

Date	Aircraft Type	Squadron	Location	Casualties
21 December 1940	Blenheim, L9415	272 Sqdn, Aldergrove	Slidrum, near Buncrana	No fatalities
4 January 1941	Whitley, T4168	502 Sqdn, Aldergrove	Glenard, near Buncrana	Three fatalities
10 April 1941	Lerwick, L7267	201 Sqdn, Lough Erne	Bundoran Strand	All safe, took off
11 April 1941	Wellington, W5653	221 Sqdn, Limavady	Close to Fort Dunree	Six fatalities
27 August 1941	Fulmar	804 FAA/RN, Cat. ship *Ariguani*	Tramore Strand	All safe, took off
2 October 1941	Hampden, AD768	106 Sqdn, Conningsby, Lincolnshire	Glendowan Mountains	All safe
30 November 1941	Spitfire, P8074	133 Eagle Sqdn, Eglinton	Glenshinny	Pilot safe
16 December 1941	Spitfire, P8267	4 ADF Ferry flight	Maghermore Strand	Pilot safe
21 December 1941	Martlet, AM975	881 Sqdn FAA, HMS *Illustrious*	Clogfin, close to Carrigans	Pilot safe
6 February 1942	Sunderland, W3977	201 Sqdn	In sea off Dunmore Head	12 fatalities
6 March 1942	Catalina, VA721	ATFERO	In sea off Malin Head	All safe
19 April 1942	Blenheim, N3533	143 Sqdn, Aldergrove	Three miles from Buncrana	All safe
4 May 1942	Hudson, FH376	ATFERO	Close to Ballyliffen	All safe, took off
16 June 1942	Hudson	ATFERO	Hill Strand, Dunfanaghy	All safe, took off
17 July 1942	Beaufort, N1063	5 OTU, Long Kesh	Ballyness Strand	All safe
25 July 1942	Lysander, 2745	Valley, Wales	Ballyliffen Strand	All safe, took off
13 August 1942	Liberator, LV341	120 Sqdn, Ballykelly	In sea 37 miles off Tory Island	All killed
12 September 1942	Wellington	7 OTU, Limavady	Lough Foyle	All safe
29 October 1942	Hampden	5 OTU, Long Kesh	Close to Ballybofey	All safe
26 November 1942	Spitfire, AD116	501 Sqdn, Ballyhalbert	Stranorlar	Pilot safe
2 January 1943	Wellington, HX467	7 OTU, Limavady	Lough Foyle	Six fatalities
9 January 1943	P-38 Lightning, AF4212802 – being shipped		Moville	n/a
27 February 1943	Wellington, HX737	7 OTU, Limavady	Near Falcarragh	All killed
18 March, 1943	Liberator, FK222	86 Sqdn, Aldergrove	Tallan Strand, Fort Lenan	No casualties
10 May 1943	B-17 Fortress	USAAF Ferry Flight	Portnablagh	No casualties
10 July 1943	Unknown RAF aircraft	In sea off Innishowen Head	Crew picked up by RN destroyer	
16 August 1943	Beaufort	FAA	Near Falcarragh	No casualties
14 December 1943	Harrow	271 Sqdn, Eglinton	Near Moville	Four fatalities
17 December 1943	Martlet	FAA	St Johnston	No casualties
23 January 1944	Halifax LK714	518 Sqdn, Tiree, Scotland	Tullan Strand, Bundoran	Eight fatalities
31 January 1944	Sunderland, DW110	228 Sqdn, Pembroke Dock, Wales	Blue Stack Mountains	All killed
20 February 1944	B-17 Fortress	USAAF Ferry flight	Killybegs, Donegal Bay	No casualties
19 June 1944	Liberator, FL989	59 Sqdn, Ballykelly	Glengad Head	Eight fatalities
19 June 1944	Liberator, FL990	59 Sqdn, Ballykelly	Shrove Hill	Eight fatalities
19 June 1944	Liberator	USAAF Ferry flight	Near Ballyshannon	Two fatalities
13 July 1944	Liberator, BZ910	120 Sqdn, Ballykelly	In sea close to Inishtrahul	Four fatalities, five survivors
12 August 1944	Sunderland, NJ175	422 Sqdn RCAF Castle Archdale	Near Belleek	Three fatalities
6 September 1944	Sunderland	423 Sqdn RCAF Castle Archdale	Donegal Bay	Nine killed
10 February 1945	Liberator, KK295	45 Group Ferry flight	Portsalon	No casualties
14 March 1945	Sunderland, ML743	201 Sqdn, Castle Archdale	Near Killybegs	12 fatalities
14 March 1945	Hurricane, PZ 774	1402 Sqdn Met flight, Aldergrove	Near Moville	No casualties
24 June 1945	Martlet, JV176	891 Sqdn FAA, Eglinton	Greencastle	No casualties

The Corridor

It has become commonplace to refer to the Republic of Ireland as 'Southern Ireland' to distinguish the state from Northern Ireland, yet the most northerly county in the whole island is Donegal, an area whose rugged coastline and picturesque mountains is a mecca for tourists. Several years before the outbreak of World War Two, the RAF had conducted a thorough survey of the approaches to the beautiful area surrounding Donegal and Lough Erne in County Fermanagh across the border. The Air Ministry visualised flying-boat bases on the Lough with a wide flight path between the border town of Belleek and the yet to be built aerodrome at Long Kesh – later to become the site of the Maze prison. The implications of overflying County Donegal in order to shorten the routes to the Atlantic and the North Channel did not, at the time, enter into the plan.

At the outbreak of war, the British envoy, Sir John Maffey, suggested joint Anglo-Irish patrols of the ocean. De Valera rejected the idea but said that when Irish patrols located any submarines, information of their whereabouts would be broadcast. "Not to you especially," Dev added, "your Admiralty must pick it up. We shall wireless it to the world. I will tell the German minister of our intention to do this." Subsequently, the British asked that such messages be sent in code, and this was done – a curious one-sided arrangement?

The question of overflying now arose when the British representative asked that patrol aircraft be permitted to get out to the Atlantic without adding to their journeys by flying a dogleg course. It is sometimes said that a certain menacing note underlay this request. Whatever the truth of this, agreement was reached: an airway extending from the cliffs that fringed the north side of Donegal Bay to Innishmurray, 25 miles to the south, was granted. Certain restrictions were laid down: no publicity by the Allies; aircraft to fly as high as possible over Irish territory; no overflying the army base at Finner and the coastal defence forts on either side of Lough Swilly. The forts in question were Fort Dunree and Fort Lenan, large fortifications in black stone and concrete which guarded the deep anchorage of the lough – which is actually a fjord.

For its part, the Air Ministry cautioned, "Personnel sent to Lough Erne bases are warned not to discuss with local inhabitants the air routes arranged." It further requested that the lookout posts flanking Donegal Bay should cease widespread reporting of movements of aircraft flying to and from Lough Erne and over the bay. This also was agreed. Of course, possession of the airspace was nine tenths of the law, and Éire had no real means of intercepting and forcing down aircraft – though it could cause difficulties to occur elsewhere.

Eventually, the stricture about keeping exactly to the corridor was more honoured in the breach than the observance as aircrews and the Irish ground troops exchanged friendly greetings. One story related by a young officer of the time recreates the ambience:

I was visiting a Lookout Post in County Sligo for the first time when I thought I saw the grey cigar shape of a U-boat through the drifting fog and exclaimed to the sergeant: "My God, man, there's a submarine on the surface charging its batteries; can't you hear the engines? Put me through immediately to Central Intelligence." He responded: "Indeed I will not, sir, for that thing below there is the Bonmore Rock which hasn't moved in the 40 years I've known it. And as for the engines, them's the shaggin' Coastal Command planes that are always flying over here, instead of sticking to the corridor we gave them over Donegal."

Later, the 'free zone' agreement was extended to give a clear run to Allied aircraft going northwards into the Atlantic by allowing them to transit over the Innishowen

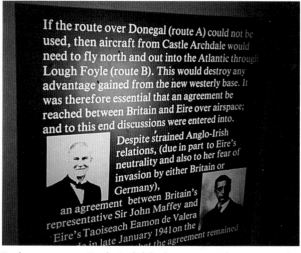

In the museum at Castle Archdale there is an explanatory panel giving details of 'The Donegal Corridor'.

Peter Brown

Peninsula rather than diverting over County Tyrone. This facility reduced two-way flights by 160 miles for aircraft covering the area west of the North Channel. Both these concessions saved thousands of airmiles, time, fuel and crew fatigue.

Shortly after the first 'Donegal Corridor' was conceded a further urgent request came from the Air Ministry to allow its armed drifter, *Robert H Hastie*, to be based at Killybegs. The vessel had been used during World War One and was now re-hired to carry out sea/air rescue work (SAR). The Air Ministry gave assurances that the vessel would be manned by civilian crews, and would fly the Merchant Navy's 'red duster'. Its two 12-pounder guns were removed, but it retained two machine-guns, along with rifles, ammunition and grenades – obviously it had to have some protection against enemy action.

The captain was informed that if it was necessary to land RAF or other uniformed personnel in Éire, they should be dressed in seamen's kit to avoid internment. The secret nature of all these arrangements was highlighted when an RAF officer visited Killybegs and met the Western Command Intelligence Officer. He was greeted "with far less cordiality than usual" because the IO had yet to be informed by his own government about the vessel. It was a case of the 'need to know' factor being carried to extremes - a major public relations clanger had been dropped. The vessel gained more publicity for its moonlighting than it did from rescuing downed airmen. The captain and his crew were engaged in smuggling between Killybegs and their depot at Derry city and were duly arraigned in court, found guilty, and fined. All, except two, were dismissed and a new crew was signed on. Killybegs at the time had taken on a 'battleship grey' aspect because the original crew had provided hard-to-get paint to all comers, for a consideration.

All these exceedingly un-neutral arrangements were not disclosed by the Irish government until several years after the war had ended. Certainly an English wartime visitor held a different view. While travelling by bus in Donegal he overheard a couple of men conversing in Gaelic, the county being a stronghold of the Irish language. He was outraged and reported that the county was overrun by Germans.

The only foreigners in Donegal were those unfortunate enough to crash-land on its wild and mountainous terrain.

A Donegal girl on board the SAR vessel *Robert H Hastie* wears the captain's cap and binoculars.

Sergeant Pilot Denis Briggs, who was later commissioned, was the pilot of the unloved Lerwick flying boat which ran out of fuel off Donegal on 10 April 1941. The Lerwick was unstable in the air and on the water and this particular aircraft subsequently sank off the Welsh coast, luckily with no loss of life. A Lerwick pilot once remarked, "One minute flying, the next minute swimming!". The type was withdrawn the month following the Donegal incident.

James Stewart

There were tragedies aplenty, none more so than on the 19 June 1944, whose misty morning heralded a black day for Coastal Command's 59 Squadron stationed at Ballykelly. Before the day was very old, one of its Liberators (FL989) had taken off on a routine patrol which ended within ten minutes when the aircraft flew into Glengad Head killing all eight crew members. The supposition was that a down draught from the mountains had somehow contributed to the accident. The Irish Army informed the Squadron when the crash site was established.

Unaware of what had happened, the crew of Liberator (FL990) took off, only to impact within minutes on the high ground beside Lough Foyle - again the complete crew were killed. An Irish Army unit was once more on the scene but like the earlier event there was little it could do. Before the troops arrived a young man, named Kearney, had pulled one of the crew from the burning wreck, despite intense heat and exploding ammunition - but the airman died shortly afterwards. The RAF acknowledged the local man's bravery by presenting him with a silver cigarette case.

Concurrently in south Donegal, a USAAF Liberator coming in on a ferry flight crashed near Ballyshannon but luckily with only two fatalities.

There were lighter moments in Donegal too. When Sergeant Pilot Guy Fowler of the RCAF was ferrying a reconditioned Spitfire from the Isle of Man to Limavady in Northern Ireland, he decided he would cut some corners to shorten the journey. His decision meant that he found himself with less than five gallons in his tanks and no idea of where he was. His actual position was over Clogher and the wide strand at Maghermore offered him a place to reflect on his navigation. The LDF were in time to stop him destroying some documents and damaging the Spitfire. His explanation for his presence was somewhat contradictory but he did say that his parents were in the United Kingdom; that he wanted to give them a good Christmas; and that he had spent all his money buying hams for the festive season. He came across to his captors as a ham actor with a fishy tale!

5 Hurricanes, Wimpys and Whitleys

In 1940, the Irish High Commissioner in London had raised a purchase order for sufficient aircraft to train a modern fighter squadron. The request was for 10 North American Harvard advanced trainers, 3 Fairey target tugs, and 13 Hurricane fighters. The order had the backing of the British representative in Dublin and his Air Attaché. By now a very good relationship had been established between the Air Corps and RAF, which the latter was keen to foster. However, the request to release these aircraft now came up against the two factors which governed all orders from Éire: availability in terms of Britain's own requirements, and a more significant stumbling block – politics. The British attitude was basically to provide minimum supplies to Éire whilst seeking maximum political and military gain. The RAF felt that it could supply six early-mark Hurricanes or Spitfires, with more to follow, but this was personally vetoed by Winston Churchill. But a trio of Hurricanes was eventually acquired without the Prime Minister's approval.

By the end of September, the intensity of the Battle of Britain was waning and the last Sunday of the month was a particularly quiet day. Nevertheless, three squadrons from the Luftwaffe's KG55 set off at six o'clock in the evening to bomb Merseyside. To avoid recent heavy losses, the formation hugged the east coast of Ireland, but radar in Cornwall spotted them. Eleven Hurricanes from 49 Squadron RAF took off from Pembrey to intercept. Two of the German squadrons eluded the British fighters, but against the setting sun the Hurricanes spotted the nine Heinkels of the 9th Squadron. Pilot Officers Mayhew and Nelson-Edwards brought down one of the intruders (a Heinkel 111-P, coded GI+DT) and two more were damaged and turned for home. On the debit side, Nelson-Edwards and George Peters were brought down by the well-directed crossfire of the Heinkels. Nelson-Edwards was picked out of the sea, his only injury being a bullet through the welt of his new bespoke shoes, but Peters was drowned. Paul Mayhew, still 'searching for trade', stayed out too long and, with fuel exhausted, landed in County Waterford. He became the first British airman to be interned. His

Pilot Officer Paul Mayhew had just shot down a Heinkel in this Hurricane when, with fuel exhausted, he landed in County Waterford. He became the first British internee at the Curragh and his aircraft was the first Hurricane to enter service with the *Air Corps*

Canadian-built aircraft Z2832, coded GZ-M, became the first Hurricane in the Irish Air Corps, being given the serial number, after repairs, of No 93.

Unlike the coastal airmen who had previously alighted in Irish waters, Mayhew had just shot down an adversary – the bodies of its crew were found on the coast. (This was an undeniable operational act, and in addition, half a dozen German airmen on a met flight had already been interned, hence neutrality could not now be stretched to allow Mayhew to be sent home.) His aircraft, in the words of the salvage team, became "The First of Our Few!"

Some months later a pair of Free French pilots from 32 Squadron, which had just moved to Anglesey in Wales, took off in their Hurricanes to intercept a Heinkel He 111 H-3 (coded '4T+JH') of l/Wekusta 51. One pilot, Maurice Remy, damaged it heavily before losing it in cloud. Though the men at Lookout Post 13 spotted the Hurricanes, they could not see the Heinkel due to scattered clouds and ground haze, and though the airborne pair searched for one-and-a-half hours, neither could they. Remy, unaware that his quarry had crashed in County Waterford, continued to search for it but ran out of fuel and was forced to make a landing.

Left: The German pilot of 4T+JH was trying to get down into a field when the gunner's gondola under his aircraft struck this huge boulder. Those of the crew who had not been killed in action died when the aircraft exploded in flames. The incident was viewed by troops who were exercising nearby, many of whom confirmed that Remy's companion came in and sprayed the wreckage with machine-gun fire. This incensed the troops who returned fire with their Bren guns. No mention of this incident has ever appeared in any official report nor indeed were two further occurrences where it was German aircraft that the Irish infantry fired on.

PJ Cummins

Right: The wreckage of He 111 H-3 (coded 4T+JH) of Wekusta 51, which was shot down near Churchtown Carnsore Point, County Wexford, by Free French pilot Maurice Remy of 32 Squadron RAF. The Frenchman's Hurricane ran out of fuel and he made a forced landing. His aircraft, duly repaired, became No 94 of the Irish Air Corps. Some of the German crew were killed in the action and the rest died when their aircraft exploded in flames.

James Joy

Meanwhile, as the Frenchman searched, the German pilot was desperately seeking a landfall. Undoubtedly he had dead and wounded aboard. He came in at sea level but the gondola of his plane struck a large boulder before he could reach his selected field. The plane disintegrated and burst into flames. The second Frenchman sprayed the wreckage with gunfire; infantry who were exercising in the area, infuriated by this ungallant act, opened fire with Bren guns. There is no official record of this response, but neither is there one covering an occasion, early in the war, when another infantry battalion opened fire with Vickers guns on a German aircraft which was suspected of killing three civilians.

From his home in Kilmacthomas, Richard Boyle saw the French man's Hurricane approaching. It flew in circles as the pilot searched for a suitable landing place, "in a field that did not have cattle grazing" as he later told the Boyle family. Five minutes later, with the undercarriage retracted, F/O Remy 'belly landed' in a corner of a large uncut meadow. The aircraft skidded through the high grass and eventually came to an abrupt stop on a slight slope in the field. Shaken but unhurt, the pilot climbed out of the fighter's cockpit and, thinking he was in Wales, made no attempt to destroy the aircraft or its contents.

Above: All who perished in the crashes were buried with dignity and full military honours. At the graveside of six Luftwaffe victims, the First Secretary of the German Legation in Dublin, holding a wreath, gives the Nazi salute, while the Irish military render more formal honours.

John Scanlon

Left: This virtually undamaged Hurricane Mk IIB is in a field in County Meath. It became the Air Corps' No 95 though its serial number, Z5070, had not yet been applied to its fuselage.

Norman Coffey

A nearby group of county council workers, members of the LDF, surrounded the Hurricane and took the pilot into custody as, having found out that he was in neutral Ireland, he attempted to return to the plane. The Boyle family served him a meal while a military escort was awaited. The Frenchman wanted to know the name of "the large town which he had seen prior to landing" and was told that it was Waterford city. In the afternoon, Remy was transferred to B Camp at the Curragh, while his Mk IIA Hurricane was transferred to Baldonnel. Here it was evident that its undersides had been protected by exterior fuel tanks which took the impact. Only these, the aircrew blades, air intake and radiators were damaged. Soon the aircraft, now No 94, was flying again.

A few months after Remy's arrival, an undamaged Mk IIB arrived in a field in County Meath. The young sergeant-pilot was from Ontario and had arrived in the UK two months earlier. On the morning of his Irish visitation he had taken off from Wilmslow, Cheshire, with three other Hurricanes on a ferry flight to Abbotsinch in Scotland (later to become Glasgow Airport for a time). However, fog enveloped the formation and his radio went unserviceable, causing him to lose his bearings. Eventually, with fuel running low, he decided to make a landing as soon as possible and chose a field known locally as the 'Cemetery Field'. It might not have been the best of omens, but he made a perfect wheels-down landing. The Mk IIB's normal armament was a dozen Browning machine-guns but two had been deleted from this ferry flight to give extra fuel capacity. The pilot was entertained by a lady who ran a stud farm close to the landing site. She had ambitions to spirit him across the border, some 70 miles away, but this plan was thwarted and Z5070 became No 95.

By 1942, therefore, three Hurricanes were in the hands of the Air Corps, having been purchased, together with spares – the pair of Mk IIs at a bargain price of £7,200. Despite these 'windfalls', supply problems had come to such a pitch that by mid-year the Army General Staff decided to disband

the air arm and redeploy its personnel in a purely ground role. All ranks had already had infantry training and, indeed, an Air Corps AA battery was already in being. The RAF was somewhat taken aback by this development. The existence of a friendly and 'flying' Air Corps was proving of great value in terms of the highly un-neutral services it was rendering. Further consultations took place and, as a result, an RAF instructor, F/L Donal West, arrived at Baldonnel in July 1942 in a Miles Master Mk III trainer to conduct a conversion course in preparation for the release of more Hurricanes. Perhaps he was selected because his mother was from County Louth where her own mother still resided.

'Don' West's aircraft was finished in British civilian markings (G-AGEK), with Irish colours added to its undersurfaces. In this unique livery, the aircraft went into service at Baldonnel. Earlier, West had participated in the Battle of Britain, flying Hurricanes and the difficult Defiant. It was in a Defiant of 256 (Night-Fighter) Squadron, six months after the end of the Battle, that he downed a Ju 88 and was well on his way to getting a DFC. Even before West's arrival, the ground crews had made themselves conversant with the Hurricanes, since they were direct descendants of the Kestrel-engined Hawker Hind and had armament and cockpit layout similar to the Gladiators.

In July 1943, four Mk Is were acquired in exchange for the two 'windfall' cannon-equipped Hurricane Mk IIs. In addition to these a premium of £4,000 and a large amount

Flight Lieutenant Don West demonstrates the first Hurricane 'windfall'.

Flight Lieutenant Don West (second from left) is seen with some of his pupils gathered around 'windfall' Hurricane No 95 in the summer of 1942.

The 'windfall' Fairey Battle arrived in time to replace the DH 84 target tug which had crashed at Baldonnel.

Jim Watkins

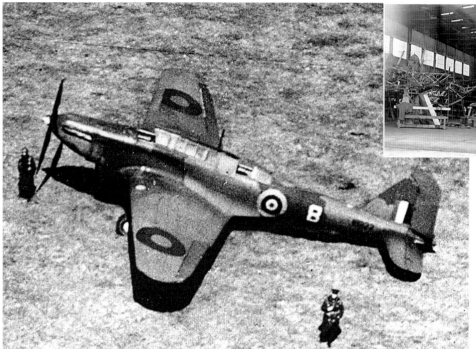

The Fairey Battle 'windfall' in Air Corps colours at Collinstown/Dublin Airport.

Above: A Hurricane being rebuilt at Baldonnel.
Cdt J Ryan

Left: The Fairey Battle target-tug which landed on Waterford racecourse.

of salvaged equipment were requested. Eleven more Hurricane Mk Is were delivered, starting in July 1943, with the last batch arriving in March 1944. They were a motley collection, all having seen varying service use with the RAF. The Air Corps could now boast a proper fighter squadron. The 1st Fighter Squadron (including No 93) never suffered a serious accident before the Hurricanes (six of the cannon-type were added in 1945) were withdrawn from service in 1946.

In the spring of 1941 the Air Corps suffered the loss of its one and only target tug, a DH Dragon 84 whose pilot had attempted to take off with locked controls. He was henceforth known as 'George' because he had slain the 'dragon'. Luckily, a few weeks later, a more modern replacement appeared when a Fairey Battle TT (V1222) arrived intact on Waterford racecourse from No 4 Bombing and Gunnery School at West Freugh in Scotland. At its controls was Kasimierz Baranowski, an ex-member of the Polish air force who had come to the RAF via the French air force whose uniform he was wearing. His English was poor and initially he blamed his landing on fuel shortage but later put his arrival down to instrumentation problems. Not only did he deliver a target tug but also a full complement of targets. The facility for air-to-air firing practice was now restored to the nascent Hurricane squadron and other aircraft, and the AA batteries too had something to shoot at.

Though it was an ill wind for the RAF men involved, it surely blew some good for the Air Corps in the shape of these 'windfalls' – four aircraft from the original shopping list of November 1940.

Rineanna 1943: a full complement of Hurricanes now equip the Air Corps' Fighter Squadron.

Captain A Quigley

Wimpys

Popeye, the cartoon character, had a hamburger-munching pal Wellington Wimpy, so it was natural to call the Wellington bomber the 'Wimpy'. It had a unique criss-cross geodetic structure which could absorb a considerable amount of damage. At the controls of one such Wimpy, T2506 of 103 Squadron, Canadian pilot Ralph Keefer was not best pleased with his new tail gunner, Sgt Alex Virtue, as they set out to bomb Frankfurt. As they prepared to take off the gunner told the pilot that he had not been able to install his guns properly and there was a delay while an armourer remedied the problem. This delay meant that 'C for Charlie' would be late over the target, by which time the German defences would be on the alert. Sergeant Virtue, having already been torn off a strip, was reluctant to mention that, as he had swung his turret back to the take-off position, one of its doors jammed open: the guns were inoperable and he was half frozen during the next eight hours.

When returning from Frankfurt, Keefer's fourteenth mission, he wondered if he would survive the remaining six trips which would complete his operational tour in Bomber Command. This figure had been reduced from 30 to 25 and then to 20 because of heavy losses. The odds were stacked against C for Charlie getting home safely: it had suffered complete radio failure, the navigator was suffering from concussion caused by flak and there was severe icing. Completely off track, the aircraft was now over the west coast of Ireland and when this was established a course was set for Northern Ireland, but fuel was down to a minimum. The crew were ordered to assemble for'ard for a mass bale-out, but Virtue plaintively announced that he was jammed

in his turret. The second pilot took over the controls while Keefer grabbed an axe and began swinging it an inch from the gunner's sheepskin jacket to cut the hinges off the aluminium door. They pushed and pulled until Virtue was able to join the others.

Ralph Keefer tells his own story:

"Mayday, mayday . . . C for Charlie C2506 . . . Over." I switched to Receive, listening and straining to hear, but there was nothing. I pulled the controls back and climbed a little higher to 8,000 feet. I watched Jack Calder, the other pilot, shaking hands and back-slapping each crew member as he lined them all up, ready to jump 'in a stick' – it was four o'clock in the morning.

The engines began to splutter and I cut the master switches and gave the order "bale out." I turned the bomber 90 degrees to the left, hoping she would fly across the coastline and out to sea. I set all the trimming

The central section of a Wimpy is moved by a mobile crane.

Sean O'Foghlu

tabs so she was gliding nicely and climbed out of the pilot's seat. Jack Calder had launched the four sergeants in quick succession and dived through the open entrance door as I stood by the control column and checked the instruments again. The altimeter read 6,000 feet, but one wing had dropped and she was starting to turn slightly to the right, probably due to the shift in weight. I knew I had plenty of time so I climbed back, plugged in my headset as I corrected the turn and changed the trim. I clicked on the switch to transmit and gave it one more try. "Mayday … Mayday … Crew Baling Out … Mayday C for Charlie 2506." For a few seconds I listened intently, but could only hear the rushing wind. I jerked out the plug and climbed down again from the seat watching the altimeter winding down towards 4,000 feet. I crouched down at the edge of the open door in the bottom of the aircraft and rolled out head first.

A sudden rush of cold air hit my face and I saw the empty plane gliding ahead of me and disappearing into the night. My gloved hand was firmly on the rip cord and as I pulled I thought, I hope to God this thing works. There was no time to ponder, within a second there was a severe jerk and I felt myself hanging with my 200 lbs supported on two straps running between my legs.

Suddenly all the noise had stopped and I was motionless in complete solitude on this dark and eerie night. I grabbed the shroud cords over my head and pulled to re-distribute some of my weight. It was so calm and still; the coastline had disappeared and it was becoming blacker and blacker. I seemed to be hanging there for quite a long time, but it was probably only about three minutes. The darkness thickened and I now had a feeling of movement and that the earth was approaching. I landed and lay still, and then my mind started to twist and turn and to fret and to worry and the remorse began to set in. What a nightmare! What a mess!

Hence one of His Majesty's operational aircraft was written off and six airmen with a year's expensive training were interned in a neutral country. All the crew were rounded up from their widespread landfalls by the Gardai, commanded by their Superintendent – the father of the present writer. The airmen and their crewless Wimpy had come down in the west of Clare, the latter just missing a farmhouse. The crew were interned but Alex Virtue's troubles were not over: while in the Curragh he was diagnosed with cancer and given the best possible medical attention before being repatriated; unfortunately he succumbed to the disease.

Exactly a year later, Sgt George Slater, the W/Op on Wellington Mk BIII of 427 (Lion) Squadron RCAF, unlike Keefer in the previous incident, had no qualms about destroying one of His Majesty's aircraft! His

On the right is this Wellington's rear gunner, Sgt Alex Virtue, who parachuted with his crew on returning from a raid over Germany. When interned in the Curragh Camp the unfortunate Virtue contracted cancer and subsequently died in England.
M Moloney

Left: This Wellington of 221 Squadron is believed to be W5653 which crashed into the Urris Hills, Co Donegal, on the 11 April 1941.

Below: This model by William Walsh represents the Wimpy which, on a bright moonlit night, was forced to land near Waterford city. One of the crew, Sgt George Slater, had no qualms in destroying 'S for Sugar', leaving just the rear turret intact.

Wimpy, 'S for Sugar' (Z1676, coded ZL-S), had an all-British crew because of an initial shortage of Canadians. In company with eight other Wellingtons, S-Sugar had bombed the U-boat pens at Lorient and on its way back its navigator passed out due to lack of oxygen. Sergeant Slater couldn't raise anyone to get a fix and after cruising around for an hour, S-Sugar, unbeknownst to its crew, was over Waterford which was clearly visible on a bright moonlit night. The pilot picked out a field and, with his landing gear brushing the tops of trees and the roof of a house, he accomplished a heavy landing – so heavy in fact that the rear gunner was thrown out but not injured. The Wellington was brought up short by a fence, which removed its undercarriage, and by a tree which caused it to swing through a right angle into the adjacent field.

There was an agreement between the British and Irish governments which guaranteed the salvage and return to the border of all crashed RAF aircraft. Unaware of his neutral location, Sgt Slater activated the explosive device which set the Wimpy on fire and everything was burnt, with the exception of the rear turret which was untouched.

Several months later another Wimpy Mk BIII (X3568, coded ZL-T) had been battering the U-boat pens at St Nazaire but strayed off course on its return and eventually the crew had to bale out. The Wimpy went down vertically into a ploughed field in County Roscommon while the crew, a Welsh pilot and five Canadians, were allowed back to rejoin their squadron.

The Local Defence Force, nowadays affectionately referred to as 'Dad's Army', was often the first official body to greet unexpected visitors. One of its members recalls:

The first, and only, time I ever saw a Wellington bomber was when one came down close to where I was working on a Saturday morning – the five-day week had not yet arrived. I got a phone call to muster with my unit and I had just time to put on my uniform, get my rifle, and rush to where the Wellington had come down in a very long, level field which had been used for hare coursing. The aircraft used up all its length and had to make a sharp turn to avoid a quarry. We arrested the uninjured crew – it was all very courteous – and I put an armed guard around the aircraft. I must admit that I felt a bit uncomfortable every time I heard creaks and noises coming from the plane, and I was worried lest it explode as a lot of petrol was leaking from its tanks. Luckily it had been on a training flight and there were no bombs or ammunition on board. Late in the evening we were relieved by an army unit but were now detailed to guard the crew who had been put up in the Imperial Hotel.

After they had a meal we invited them to have a beer or two, an invitation which they readily accepted. Around

For the family album: two infantrymen salute smartly against a background of a Wellington which crashed in Charleville, County Cork. The army had taken over from the LDF men who had been detailed to guard the crew in a local hotel.

midnight I found that I had to go to the toilet when I heard one my chaps shout "Stop! – Don't move," followed by the unmistakable sound of a round being put up the spout. I saw one of the Englishmen framed in the bedroom door, motionless, looking at the sentry's rifle. It turned out that he had a similar call of nature and was frozen by the alert part-time warrior. In the morning farewells were said and the convoy left for the Curragh: I hope they survived the war and now enjoy a comfortable old age.

They did – like their Wimpy, they could take a lot of punishment.

Whitleys

To a certain Sgt George V Jefferson of Belfast fell the dubious distinction of being the only Irishman to be interned at the Curragh – where he spent almost three years. Jefferson had been one of the crew of a Whitley Mk V T4168 of 502 Squadron, originally the 'local' Royal Auxiliary Air Force unit. His aircraft crashed at Cockavenny in Donegal on 24 January 1941 when it was returning from an anti U-boat patrol. The navigator could not find Innistrahul Lighthouse, the essential beacon for getting into Limavady – much to the chagrin of his pilot, Les Ward. The Whitley flew on across to the Scottish coast and then re-crossed and eventually found Lough Foyle, but overshot the base. As a second approach was being

attempted, the effect of the constant searching kicked in – the tanks were dry. Baling out into the darkness was the only option. Ward and Jefferson landed safely – the former close to where his aircraft had dived into a bog, missing a school. The other three came down in Lough Foyle and disappeared into its waters.

Ward made his way to a nearby house and, when the LDF arrived, the owner, a former Royal Irish Constabulary man, would not hand his visitor over until the regular army and Gardai arrived. While being refreshed in Lough Swilly Hotel, Ward was incredulous that Éire remained neutral, saying that he had often aided Irish ships. Nor could he understand why the Hudson which had come down earlier had not been destroyed by its crew. His overriding concern was about the remains of his aircraft which was fitted with a new American bombsight and he asked if he could visit the crash to destroy it – but this was denied. Four Royal Navy officers now appeared at the local Garda station and shared his apprehension – which turned out to be unnecessary as the explosion of part of the bomb load had scattered the wreckage over a wide area. The local postman described the aftermath of the accident thus:

When the remaining bombs were detonated, the little damage this caused in the vicinity provoked no claim. However, a man whose abode was almost a mile away, claimed that his ceiling had been cracked by the explosions – he received short shrift.

Wing Commander Brian Corry, a Belfast man, who as a Squadron Leader parachuted from his Whitley into County Donegal. The first person he met was a lady who had looked after him as a child!
R Corry

The controlled explosion appeared to have disposed of all danger at the time. Four machine-guns were retrieved but a single Vickers K-gun could not be found until, after ten days' thorough searching, it was eventually located in the possession of the Local Security Force whose members were supposed to be an unarmed body! These blue-clad 'warriors' (especially the veterans of the War of Independence) had warlike ambitions and had also appropriated two drums of ammunition, Verey pistols, and signal cartridges – which they reluctantly handed in. Of the many thousands of rounds on the aircraft, only 3,500 were recovered. The scrupulous nature of searches to recover all arms, explosives and ammunition was vital because of the real danger of these falling into the more formidable hands of the IRA.

A soldier unofficially 'liberated' several rounds to improve his musketry skills and was duly censured, while a schoolgirl threw some into the classroom fireplace to enliven lessons – luckily without casualties. Such antics were repeated some 50 years later when a group of aviation enthusiasts investigated the crash site and unearthed a bomb which had not been detonated. Foolishly they hauled this lethal load up to ground level on a rope before alerting the authorities.

Seven weeks after the incident of 24 January 1941, a third Coastal Command Whitley (P5045) crashed into Galway Bay at night when an engine problem developed. Its radio was unserviceable so the operator was unable to raise his base to report the situation. The tips of one propeller had sheared off and caused its engine to vibrate violently before it was shut down. The aircraft lost height so the dud engine was opened up again but nearly shook itself out of its bearings and caused the wing to smoke. The pilot ordered his three men to bale out, and as one of them descended he saw that the Whitley was enveloped in flames before it hit the sea.

Two of the crew landed safely, close to Galway city, while the pilot's body was washed up on the north side of the bay – he had drowned, though wearing his Mae West life jacket. After an inquest, which returned a verdict of death from exposure, the unfortunate young man, his coffin draped in the Union Flag, was reverentially sent back to Northern Ireland. A detachment of the 1st (Irish-speaking) Infantry Battalion, headed by its pipe band playing a lament, escorted the cortege to the county boundary.

Of the fourth member of the crew nothing was seen, though some distance from where the two survivors had landed on the seashore, a parachute harness was discovered. A young girl on her way to school reported that she had seen a tall stranger in a grey jacket with an LDF-type cap. However, the local Gardai were sceptical, saying that the girl was one of a family of 'notorious liars'! Nevertheless, the owner of the parachute harness was never located and the supposition is that he had attempted to walk to freedom, but had fallen into a deep water-filled boghole and

drowned. The matter remains a mystery to this day.

There was another unsolved mystery. When Whitley survivors from two separate incidents were discussing their aircraft at the Curragh, one mentioned that in his squadron 12 aircraft had been afflicted by propeller tips shearing off and sabotage was suspected. The problem had apparently been noted when the type was in service with Bomber Command but no one seems to have passed this information on when the Whitleys were transferred to Coastal Command. No definite conclusion seems to have been reached on the subject.

Brian Corry, a pre-war local pilot of 502 Squadron and now a Squadron Leader, was returning from a patrol in another Whitley V (Z6553) when, 300 miles off the Donegal coast in gale-force winds, one engine failed, forcing the aircraft down virtually to sea level. Corry managed to get it back up to 500 feet, but when he crossed the coast the other engine caught fire. Despite the low altitude, he and his crew had to bale out in total darkness. Corry landed close to Lough Melvin in Co Leitrim through which the invisible line of the border runs. He stumbled across the moors to a lone farmhouse, where he was astounded to meet the woman who had served his family as a maid when he was a little boy! When he had been well fed and warmed up she gave him a bicycle and off he headed round the lake into safe territory. While he was chatting to a member of the RUC he suddenly remembered that the then secret ASV radar could possibly still be intact in his aircraft where it had crashed in Donegal. He borrowed civilian clothes and a car, found the wreck, and demolished the radar with a brick! Thus ran the story of Corry's incursions and excursions over the border. Eventually, as a Wing Commander, and using the knowledge gained from his family 'being in timber', he collaborated with yacht designer Uffa Fox to produce the first airborne lifeboat prototype which he named after his son. These boats were often successfully dropped by parachute to aircrews adrift on the ocean.

6 'Halibags', Blenheims and Hampdens

The purpose of RAF Heavy Conversion Units was to accustom crews of twin-engined aircraft to operating four-engine types. From one of these units, 1663 HCU based near York, the all-Canadian crew of Halifax EB-134 set out on a November night in 1943 on a training flight. This flight should have turned around over Wales and returned to base. For some unfathomed reason the aircraft flew on and on across the Irish Sea until it reached the west coast. It circled Tuam in County Galway for about an hour before attempting a landing which ended up at a prehistoric ring fort – a 'fairy fort' - where it burst into flames. An inferno quickly engulfed the whole aircraft and its occupants.

A woman, then a child of eight, remembers how her father and a dozen others cycled to the crash site with buckets. Fetching water from a nearby spring, they attempted to quell the fire but before water even reached the aircraft it was evaporated by the intense heat. No children were allowed near the grisly scene on the following day when transport from RAF Aldergrove arrived to take away the minimal remains of both the aircrew and their aircraft. Years later they were told of the terrible screams of the trapped crew.

Further north, the residents and visitors at Bundoran, a popular resort on the Donegal coast, were well used to aircraft noise as planes plied 'the corridor' going to and from their Northern Ireland bases. But on a cold winter night in early 1944, the noise seemed to have a more ominous note as a Coastal Command Halifax (know as a 'Halibag') from Scotland, returning from a Met flight, came in over Tullan Strand. Two ladies from the staff of a local hotel were walking along a cliff-top known as 'Fairy Bridges'. They could just make out an aircraft skimming the waves and stood, frozen with fright, as it came directly towards them – another tragedy was surely imminent.

However, to their relief the Halifax turned sharply and flew back out to sea before it turned again on a reciprocal course and smashed into the cliff-top 50 yards ahead of the horrified pair. The plane disintegrated, its rear fuselage falling down into the freezing sea. Help was quickly on hand but there were no survivors, though most of the dead were recovered. There were eight crew members, one more than usual, a mix of RAF and RCAF personnel. The authorities across the nearby border were notified and some components were handed over, but the army buried most of

Below is the memorial unveiled on 12 September 2002 recalling the Halifax crew lost in the horrific crash at Tullan Strand, Bundoran, County Donegal, on 23 January 1944.

Joe O'Loughlin

Right, below and bottom: Halifax GLL-145 ended up beside a river in Skibbereen in County Cork (see text opposite). Men from the Air Corps were assisted by the 2nd Cyclist Squadron in dismantling it and returning it to Northern Ireland.

Michael Keating

the wreckage deep in the sand.

In the communal memory of the resort the tragedy faded with the years, though in unusual current and tidal conditions the site of the crash could sometimes be identified. A violent storm later washed away most of the access road to the beach, and mechanical diggers were brought in to reinstate it. This work disclosed the aircraft remains again – coincidentally on the fiftieth anniversary of VE Day – and revealed belts of ammunition which had somehow been missed at the time of the accident. Though obviously inert, no chances were taken and the Ordnance Corps finally disposed of this potential hazard.

A later Halifax was more fortunate when, with just enough fuel for 30 minutes left in its tanks, it put down off the Sligo coastline on 9 February 1945. The pilot hoped that his crew, equipped with life jackets and a dinghy, could make it to the shore. However, a strong sea was running and, though four of the crew managed to get into the dinghy, it was the Killybegs lifeboat which rescued all hands. The Admiralty's SAR (surface to air rescue) steam drifter *Robert H Hastie*, based at the same port, proved most unhasty in this instance. The national press was holding a golf competition nearby and captured the scene on cameras, only to have their films confiscated. That's how strictly censorship was applied.

Later still, at Skibbereen in County Cork, another 'Halibag', coded LL-145 from No 517 (Met) Squadron, managed to make a conventional wheels-down landing. It had been returning from a weather recce over the Bay of Biscay to its base at St David's when lack of fuel, caused by interminable circling to pick up its bearings, forced it down with barely ten minutes of flying time left. Some of the crew averred that they were actually on an SAR operation, but this was undoubtedly the usual 'cover story' in which they had been well rehearsed.

One airman tried to return to the aircraft for what he described as 'an item' – obviously something incriminating to do with the nature of the flight – but he was prevented. A medical officer checked out the pilot and one of his sergeants and decided that the latter should be confined to bed for two days – otherwise all were fit. The only real casualty was a two-year old heifer priced at £18 for which £30 was claimed. Perhaps the cost of burying the poor beast had been slyly added as a try-on? In the event, only the basic market price was paid.

The crew was taken to Rineanna where the local Red Cross detachment was holding a dance, and the airmen were cordially invited to attend. However, the Irish base commander, with great regret, turned down the invitation, saying that as the crew had already consumed a substantial amount of drink after their adventure, their presence might give rise to 'incidents'. The RAF men accepted his ruling, but the Red Cross girls resented it – no doubt the presence of valiant young airmen would have added greatly to their enjoyment. This Halifax, incidentally, was returned intact – in exchange for a supply of tea. Teabags for Halibags, perhaps (though the refinement of 'little perforations' had yet to appear)? Tea, loose or in bags, was very acceptable in a country which has the highest per capita consumption in the world and was then a commodity severely rationed.

Blenheims

There were two versions of the Bristol 'Blenheim': the short-nosed Mk I and the long-nosed Mk IV. The type was much liked by its pilots but this did not extend to the Mk I if flown on 'ops'.

The pilot of a Blenheim from Jubey on the Isle of Man picked a most inauspicious day, Friday 13, to run out of fuel while over Crossmolina, County Mayo. He circled the small town several times looking for a flat field – but all had been spiked by the LSF during 1940 – and he was in the stone wall terrain of Connaught. In desperation, he even

The wreck of a Blenheim which crashed outside a farm near Crossmolina, County Mayo, on Friday the 13th!

Vincent O'Callaghan

The amiable Polish sergeant-pilot Stanislau Karniewski landed his Blenheim on Clontarf Strand, north of Dublin city. He was interned and became a favourite at the Curragh.

Clontarf Yacht Club

considered landing on a road. The Blenheim was heard by the O'Malley family and the father, a member of the LDF, correctly surmised that it was in trouble. Grabbing a sheet from the clothes line, he ran to a field which was the only possible landing strip and waved vigorously. The pilot spotted his signal but came in too low over a stone wall which demolished his lowered undercarriage. There was a loud bang as the plane hit the ground and the farmer, now joined by his daughters, ran to the wreck.

Two officers were injured and a third man badly shocked. One of the girls, a member of the Red Cross, immediately put her skills into practice. The injured had lost a lot of blood and she tended them until an army unit arrived and transferred them to Castlebar Hospital. On the following day, the O'Malley family attempted to visit them but there was an armed guard which allowed no-one in. However, the gratitude and good wishes of the three men were expressed to the family by a senior RAF officer. Later, Claire

O'Malley, the first-aid expert, received the following communication from the girlfriend of the pilot, Welshman Henry Thomas:

I should like to thank you and your father for the great kindness you showed to Henry and to his companions. They were certainly very fortunate in having you at hand to help them and I have since heard from Henry just how good you were. You will be pleased to know that last week Henry was home and you can imagine how pleased we all were to see him looking quite well and very pleased with himself. He is back at his old station now. We shall never forget your kindness and, as Henry suggested when he was at home, maybe when the war is over we shall be able to visit Éire and then we shall be able to thank you and your father personally.

Another Blenheim, piloted by a Polish sergeant, came down on Clontarf Strand on the north side of Dublin.

William Duggan (apprentice class of 1937) was part of a team sent out to deal with it. Billy recollects:

> We had attempted to haul the Blenheim onto dry land, but it was submerged by the incoming tide. First though, I was given the job of cutting the damaged ailerons from the wings. Needless to say I was anxious that the fuel tanks were drained before I got to work with my torch! We opened the vents of the fuel tanks and sea water poured in, but there was no sign of fuel. I decided to go ahead while my pals were removing the propellers. Suddenly they shouted: "Petrol, Billy, petrol!" Quick as a flash I turned off the oxyacetylene gas and we all landed together on the ground and ran like hell. One of the lads said: "Jaysus, Billy, your back is all wet with petrol – you sure cut it fine!" What had happened was that overnight, water had seeped into the tanks and petrol, being lighter, floated to the top. We then waited till we were certain that all the petrol was drained off, and we moved the Blenheim from the petrol-polluted area before completing the dismantling. It's thanks to my pals and my guardian angel that I have lived to tell this tale!

The Blenheim was later towed into Dublin docks for shipping back to England.

In the UK, 'browned-off' RAF types occasionally said in jest: "To hell with all this – let's bugger off to Éire, like that bastard in the Blenheim." This was obviously a inference that a pilot had purposely dropped out of the war – but it is unlikely that it referred to the Pole whose courageous countrymen were always eager to get to grips with those who had overrun their homeland.

A more serious Blenheim crash involved an all-Canadian crew from 236 Squadron, based in south Wales. This aircraft was forced down in the sea between the Long Islands near Schull, at a point not far from where a Luftwaffe Condor had crashed earlier. The dorsal gunner of the Blenheim was Sergeant Charles Brady, of Irish descent, who takes up the story:

> Our job was to patrol convoys from the Atlantic into the Irish Sea and see them safely into port. We were flying the Blenheim Mk 4F which had four forward-firing guns. I operated a single gun in my cramped turret. The day before we ditched, one of our squadron had lost an engine, but had managed to get home safely. We were suddenly faced with the same problem when our starboard engine suddenly packed up, and a few minutes later, the port engine's oil pressure started to shoot up – we had no alternative but to ditch. Because of our location over the convoy, we were prohibited from sending out an SOS or turning on our IFF (Indicator Friend or Foe) transponder to give our position and situation.
>
> We struck the water at about 120 knots. I got out of my turret and crawled to the back where I was able to open the rear hatch and exit. Paul the pilot got out through the top, but poor Dougie Woodman, the navigator, who was in the nose and had forgotten to strap himself in, was flung right through the front of the aircraft into the water. We helped him into the dinghy which we had launched, but his bones were sticking through his knee and his chest was all caved in. Despite his injuries he asked me for a cigarette.
>
> When we had been floating around for about seven hours, we were towed ashore by the island people before a motorboat came up and brought us to the mainland. By now Dougie was in agony and he was taken straight to hospital. Paul and I were in pretty bad shape and a priest came in to comfort us, bringing a bottle of brandy and the pair of us sank the whole damn bottle which, I guess, sort of softened us up! We were interrogated, and then started off on the long ride up to the Curragh internment camp. We had hit the sea at noon that day, 23 October 1941, and as midnight struck we were lodged in the Curragh. Our pal Dougie died and was buried with great reverence and full military honours in the town of Mallow.

An observer at the funeral, then a young boy, remembers:

> The Catholic Church at the time was very strict, over strict in fact, as I will tell you. As his funeral passed through the main streets of the town with the band playing 'The Dead March', blinds were drawn and doors closed and large crowds lined the streets and followed the remains to the cemetery. The Last Post was sounded and three volleys fired over the grave.
>
> The following day at school we were addressed by the local parish priest in our assembly hall who said that all of us who had attended the Protestant burial were guilty, as Catholics, of a very serious transgression. We were all made to go to confession and afterwards the priest said that he was now prepared to accept that we were not aware of the gravity of our actions. The pressure put on people to remain well within the

Above and below: Sergeant Pilot David Sutherland was on a training flight from the Isle of Man to Northern Ireland when he became lost and force-landed his Blenheim at Jenkinstown, County Louth. He suffered a broken ankle – in contrast to the massive loss of life when a Liberator crashed in the same area.

Col C Mattimoe

parameters of their own religion can only be fully appreciated by those, like me, who lived in those far-off pre-ecumenical days – it was religious bigotry at its worst which would not be tolerated today.

I later gathered that the young airman – he was only 23 years old – was in fact a Presbyterian. He still lies in the local cemetery where his grave is beautifully tended and marked by a simple headstone bearing his name, his rank in the Royal Canadian Air Force, and the date of his death. Anniversary services are sometimes arranged in Protestant churches and are attended by people of all faiths.

The lost squadron

Before Pearl Harbor, a US general was touring British installations in the Middle East, including the base close to the Suez Canal of 108 Squadron RAF which was then equipped with Wellingtons. He sent a signal requesting an urgent delivery of 16 B-24 Liberators to this squadron from part of a contract for l'Armee de l'Air (the French air force) which had been taken over by the British Aircraft Commission. Pilots from the USAAF Air Transport Command were detailed to ferry the aircraft, one of which

was written off at a stop-over in Africa when it plunged into a construction site. There were no injuries and the next Liberator coming through picked up the grounded crew. The Americans stayed on for a month with 108 Squadron to familiarise the new owners with the Libs, one of which successfully bombed Tripoli in Libya at the beginning of the New Year.

Pearl Harbor precluded delivery of any further Liberators by the southern route, so the squadron commander requested permission to despatch a Liberator to the UK to collect more of the type which had been ferried there by the North Atlantic route. This plan sent Liberator AL-755 back to 'Blighty' with Wg Cdr JR Wells as pilot. The passenger load comprised six pilots, three navigators, three wireless operators and a fitter, together with the standard crew of six. All were naturally in high glee. They would form the nucleus of a re-equipped and revitalised squadron which would be brought up to strength at an OTU.

Not long after leaving Egypt, a recall signal was received warning of increasingly bad weather ahead. The Liberator was at the point of no return and, though the radio had packed up despite the efforts of all three wireless operators, virtually everyone was in favour of pressing on. Hours later, the Liberator was flying above solid cloud and had passed its ETA in Dorset – ditching became a distinct possibility. Wing Commander Wells cautiously brought his aircraft down through the cloud and observed a large bright glow which he was convinced emanated from Dublin city. A landing in Éire, he reckoned, would probably mean internment for all, so he altered course for an airfield in County Down, even though fuel was running very low. This change in course proved to be a fatal one when the aircraft crashed on Slieve-na-Glough, a mountain bog over 1,000 feet high. Only four of the passengers survived – the core of a new squadron had been annihilated.

A rescue team had a two-mile climb to get to the crash site which appalled its officer, Commandant Niall Harrington: "It was almost impossible to fully describe the scene – the whole area being strewn with bodies, baggage and debris." The badly injured survivors were taken across the border in an Irish ambulance but, due to the vehicle's provenance, a certain section of onlookers pelted it with stones and missiles to the extent that the patients had to be transferred to a British ambulance for the onward journey to hospital.

There was an another unpalatable sequel to the crash. A newspaper headline read: "RAF men's effects kept in Éire, says MP" Apparently, Sir Ralph Glyn had been asked by a relative of one of the dead to raise a question in the House of Commons as to why next of kin had not received the personal effects of the casualties. This was wholly untrue: the commander of the rescue team had caused all items to be collected and listed at the crash site, and everything – all luggage, parcels and loose articles – had been taken to Dundalk military barracks where they were further sorted and labelled with the owner's name and then handed over to the British authorities.

Later, when two members of the squadron returned from Egypt, they attempted to retrieve personal items and gifts belonging to four of their dead comrades, which would normally be held at the RAF repository at Ruislip. However, at this depot the two visitors were not made welcome when they discovered that nothing had been sent on to next of kin. The question that was to be raised in the House was quickly withdrawn. The Liberator accident happened on 16 March, the eve of St Patrick's Day, and coincided with another tragedy for the squadron when Liberator AL-574 crashed on take-off, far away in Egypt.

Regarding the Irish crash: there was a supposition that despite there being three radio operators on board, none was entirely proficient with the American Bendix system whose crystals were unsuitable for UK radio traffic.

The 'poor bloody infantry' were always involved in rendering help and guarding crash sites, and in the view of their officers required special foul-weather kit. Reporting on the conditions endured during the events on Slieve-na-Glough, the officer-in-charge said:

> My men were so fatigued afterwards that I ordered 72 hours' excused duty for them. Both myself and other officers engaged at previous crashes feel strongly that present field cooking equipment (camp kettle, turf and timber, and dry rations) and even the petrol field cooker, on account of its weight, is not suitable for mountain rescue work. In addition, getting to and working at a crash site in darkness, with the present issue of completely useless torches, poses the danger of stepping on bombs, and indeed of igniting petrol if matches are employed. Men need large torches that can be strapped to the head or body, leaving the hands free. They also need proper fleece-lined oil-skin coats and hats.

The foot soldiers, at the mercy of a parsimonious civil service, were never issued with this protective clothing – but they still soldiered on in their 'bull's wool' uniforms and waterproof capes.

Apropos of the hardships of the 'poor bloody infantry', in many locations the local people could not do enough to supplement the rations of soldiers and salvage teams. Sometimes there was a hidden agenda as when a good country lady baked scones and generally cosseted the troops: before they departed, however, she presented the officer-in-charge with a sizeable bill which included the item "use of the front room".

The total loss in the Liberator accident, the worst wartime one in Éire, was 18 experienced airmen: four pilots, navigators, observers, gunners and radio operators – the only survivor was one of the latter. In the summer of 1942 the Wellingtons of 108 Squadron were withdrawn from bombing and supply-dropping missions over the Balkans and in the autumn most aircrews were posted to other Wellington squadrons. Those remaining operated the Liberators as the Special Operations (Liberator) Flight until reduced to a cadre and disbanded at the end of the year. Exactly one year after the Irish accident, 108 Squadron reformed, rising from the ashes in the words of the squadron motto, "With gathering strength".

Fleet Air Arm

The Fairey Fulmar, a somewhat heavy aircraft, was easy to land on a carrier deck because of its wide undercarriage. It was the first purpose-designed UK fleet fighter – in contrast to earlier RAF machines which had been adapted for sea duty. Fulmars were also used perched on catapults in the bows of merchant ships. Merchant Ship Fighter Units (MSFUs) had been formed before there were sufficient small escort carriers to protect the Atlantic convoys. The 60-foot catapult was actuated by cordite rockets which pushed the aircraft off at 60 mph, the pilot and observer experiencing considerable longitudinal gravitational pull.

The MSFU idea had a major drawback for aircrew: once their mission was completed they would have to head for the nearest point of land or otherwise ditch their aircraft, hopefully close to a vessel. A nineteen-year-old pilot opted for the former. His merchant ship had been outward-bound in a West African convoy when, about 270 miles west of Bantry Bay, a Condor was sighted. Unfortunately his vessel had sailed before a speedier Sea Hurricane had been taken on board. Sub-Lieutenant Birrell (just promoted from midshipman, and later to reach high rank) and Leading Naval Airman Sykes were launched in a slower Fulmar (N-4072). They tracked and engaged the Condor for 15 minutes before losing it in cloud. They continued the chase until, having expended all ammunition, they began to look for the convoy, which was nowhere to be seen.

A Grumman Hellcat from 891 Squadron FAA, based at Eglinton, made a forced landing on the ninth hole of the Greencastle golf links. It was hardly damaged but the club secretary wanted compensation for the state of the green, while a nearby farmer sought recompense for his damaged corn crop. The Air Corps got the plane on to a low-loader from Northern Ireland – the man in casual sports clothes was the leader of the British team.

The Fulmar's fuel gauge showed Birrell that it was just possible to reach Ireland, and after two hours in cloud and snow, he spotted Tramore Strand in County Donegal, where he touched down at 10 knots into a 60-knot wind. He jumped out to see what was wrong with his observer who had been groaning in pain throughout the flight. It transpired that the catapult launch had shaken loose the radio batteries whose acid had sprayed onto the hands and face of the unfortunate Sykes. Revived, he helped Birrell to destroy maps and papers, difficult in the high wind. Now an extraordinary apparition appeared in the wilderness – a young man who withheld his name but said he was an Eton scholar holidaying locally. Amazingly he bore two jerrycans of fuel, the very stuff the Fulmar needed. Refuelled, the Fulmar whipped around into the wind and had barely got airborne, with waves slapping its undersides, when Gardai and cavalry scout cars appeared on the sand dunes.

The Fulmar just made it to the Fleet Air Arm (FAA) base at Eglington near Derry, which was still under construction. It landed on a freshly-made section of runway where husky navvies threatened them with picks and shovels and unprintable language – their smoothly-finished surface had been totally ploughed up. The station commander was equally rude and told the gallant airmen to clear off forthwith. They eventually managed to tell of their adventure and apologies were forthcoming from everyone.

In addition to this touch-and-go Fulmar, the FAA put in some less dramatic appearances. On 1 September, a Fairey Swordfish, the much maligned or admired 'Stringbag', put down at Gormanston. It had been on a communications flight between the Isle of Man and Northern Ireland when its crew became completely lost in bad weather. They were given some refreshments and, of course, put back on course.

Three days later, another Stringbag came down in County Sligo. This was more dramatic than the Gormanston one because it demolished a hen house, 30 chickens and half-an-acre of cabbage. More important than

Another view of the downed Grumman Hellcat from 891 Squadron.

While the Far Eastern war was still raging in July 1945, this Fairey Barracuda of the Fleet Air Arm came down in County Wexford during a training flight from its Isle of Man base. Following the attentions of an Air Corps mechanic, it departed for home and was not needed because Japan surrendered during the following month.

this demolition, inside the aircraft was a printed statement: "This aircraft is fitted with USA/Mk Two equipment and must be destroyed immediately on making a forced landing." The pilot, who said he was on air-sea rescue duties, was allowed to put in a call to the British Embassy in Dublin asking for instructions. He was told to do nothing, and in due course the ubiquitous F/L 'Rory' O'Moore arrived and arranged for the stricken Swordfish to be removed to Ballykelly. Even at this stage of the war 'radio location' or radar, though now much more mobile, was still 'a secret weapon'. The farmer was duly recompensed for his hen house, its occupants and the cabbage.

Just before Christmas 1941 a group of pilots from 881 Squadron flew a group of the new American Grumman Wildcat Mk II naval fighters off the carrier HMS *Illustrious* somewhere in the North Atlantic. The aircraft, which were known in the Royal Navy by the gentler name of 'Martlet', were to be delivered to an air base near Rugby. The pilots were to follow their leader as they were not provided with maps and the radio installations were incomplete. But the weather suddenly clamped down and in heavy cloud they lost sight of the leader.

A young New Zealander, Bruce Girdlestone, soon found himself alone and lost, so when he spotted a likely looking field he attempted a forced landing, only to have his aircraft flip over and trap him upside down in the cockpit. Three Gardai from nearby Carrigans in Donegal dug him out and when they arrived at their station, Girdlestone realised that he had all the secret performance data of his new aircraft on his person. He had had a good meal before take-off and couldn't face the thought of eating the documents. However, he managed to flush these down the lavatory

before he was transferred to the Curragh internment camp just in time to celebrate Christmas with 38 other occupants of the British compound.

In March 1944 a Seafire, LR-841, which had taken off from Ballyhalbert to search for a downed USAAF aircraft, stayed out too long over the Irish Sea and made for land at Gormanston. Its visit presaged the dozen Seafires which were to be based there a few years later, when the Air Corps acquired the naval version of the Spitfire. The FAA pilot was profuse in his thanks for the courtesy and assistance he had received before making his way back to base.

After VE Day, the Fleet Air Arm was still training for the final assault on Japan. The Royal Navy had formed a huge task force which had arrived in the Pacific – but shortly afterwards the A-bomb made it redundant. Fairey Barracuda II MD897, coded R5D, would probably have been part of this formation. In June 1945 this aircraft, which had taken off on a training flight from Ronaldsway airfield on the Isle of Man, developed a leak in the hydraulic system, causing the cockpit to fill with fumes and making breathing difficult for the three-man crew. Concurrently, trouble developed in the cooling system of the 1,640 hp Rolls-Royce Merlin 32 engine and the pilot had to force-land in a cornfield just south of Rosslare, the ferry port in County Wexford. Local Gardai and later soldiers from a military outpost guarded the aircraft. A lone Air Corps fitter arrived on the following day and carried out the necessary repairs over two days. The Barracuda took off to return to its base but encountered bad weather and had to stop for an hour at Dublin Airport en route. This Barracuda did not visit Far Eastern waters, the domain of its namesake. Shortly after this Irish visitation, VJ Day was

Full military honours being rendered during the funerals of an RAF Hampden bomber crew which had crashed near Blessington, County Wicklow.   *Irish Times*

signalled. The Barracuda was as equally unloved by the FAA as the vicious fish which gave it its name.

Hampdens

The Handley Page Hampden was, by virtue of its configuration, known variously as the 'Flying Suitcase', the 'Flying Frying Pan' and the 'Flying Tadpole' because of its deep forebody and slender tail.

Despite these soubriquets, it was the only RAF bomber in the early days of the war with a range of 2,000 miles. It could, technically at least, get to Berlin and back. On the night of 17 April 1941, a Hampden squadron, No 50, was doing just that. One of the aircraft, AD730, reached Berlin but was forced to bomb an alternative target. It turned for home, running the gauntlet of flak and had seen off at least one fighter which became evident when it reached Ireland.

Before this, and despite very poor visibility and bad weather, the aircraft reached the British coast. It was given a 'second class fix' (not 100% accurate) near Watton, but for other unknown reasons the aircraft continued flying over Wales and onwards across the Irish Sea. The pilot was an experienced pre-war Volunteer Reserve man, so conceivably he knew at this point that he was off course. He was reluctant to lose height until clear of the Welsh mountains and over the Irish Sea. Plotting in lookout posts showed that the Hampden was circling as its crew tried to determine their whereabouts. Suddenly there was silence as both fuel-starved engines stopped and the aircraft impacted on Black Hill, near the pleasant town of Blessington in County Wicklow. A shepherd discovered the plane the next afternoon with its crew lying in a bog some 150 yards away.

The brave crew were buried after a service in the local

An Irish Army guard of honour fires a farewell salute for the crew of the Hampden which crashed at Blessington.

Protestant church: all shops closed and all houses had drawn their blinds. As the cortege moved along the main street, lined with soldiers resting on reversed arms, a band played 'The Dead March'. The wife of Sir John Maffey, the British Representative to Ireland, was the daughter of Lord Beaverbrook and this accounted for one of the many wreaths bearing the legend 'From Winnie and Clemmie'. In the fullness of time a nine-foot-high stone monument was dedicated on the spot where the Hampden had come to grief.

Another Hampden incident took place just before proper salvage equipment was acquired by the Air Corps. A team was attempting to salvage an aircraft from No 5 OTU, Long Kesh, which had been out on a night-time photographic and navigational exercise. The crew of four sergeants had no idea where on earth, or in the air, they were – so they attempted a downwind landing. The

Hampden went through two hedgerows, demolished a telephone pole, and ended up in a potato patch from which it had to be removed. The team leader describes how this was accomplished:

We floundered and struggled in a sea of mud but, eventually, got most of the aircraft out on handcarts. However, the large radial engines, each weighing a ton, could not be shifted, despite the all-out efforts of all ten men. I remembered an old hypnotic trick and told the lads to place their hands on the propeller shaft, one on top of the other in rotation, and to press down hard. I then ordered "break" and "lift", whereupon all hands heaved together. This time, the engines virtually floated up onto the cart, and the team looked at each other in wild surmise and marvelled at the power of auto-suggestion!

7 Condors and Heinkels

After the fall of France, the Luftwaffe took over the aerodromes of l'Armee de l'Air (the defeated French air force) in the occupied zone. German long-range reconnaissance and weather units, whose reports were essential for forward planning, could now operate far afield into Irish airspace and beyond. Weather conditions round the island were frequently dire and made the task of fighter interception by the RAF difficult – but its pilots often got to grips with intruders.

In November 1941 a reconnaissance Ju 88 crossed the Waterford coast and calmly flew up the centre of the country towards its target, Belfast. At Ballyhalbert, 504 Squadron put up to two Spitfires to intercept. Low haze over the city prevented the Ju 88 crew from getting good pictures, though they did see a large convoy mustering there. On their return, they were spotted by fighters but escaped into cloud and got home via a reciprocal course.

In the following spring, two Spitfires from RAuxAF 452 Squadron were scrambled from the Isle of Man to attack another Ju 88. The Aussies were joined by Poles from 315 Squadron and managed to damage the Germans' port engine. In the early part of the action one of the Australians had managed to silence the rear gunner but not before his own engine had been set on fire. He baled out at 12,000 feet and was eventually picked up suffering from five hours in the water. When the Pole returned to base, he claimed the Ju 88 but he was premature because, though damaged, the aircraft was observed heading for home. Once again 'Jerry' had got away.

A fortnight later it was the Australians' turn to triumph when Beaufighters from 456 Squadron, shepherding a convoy down the Wicklow coast, tenaciously hung on to a Ju 88 as it flew enclosed in sea mist at about 100 feet. At sea level and closing to 300 yards, one Beaufighter got in some steady bursts of cannon fire, but it was not until the following day that its kill was confirmed when a trawler off the Wexford coast picked up the crew of a Heinkel from KG-40. The pilots of 504 Squadron, nicely poised on the Strangford peninsula, were determined to put a stop to

The photos on pages 64–5 recount the mission of Heinkel 111 1G+LH (narrative on page 65). Here we have Lieutenant George Fleischmann, the cameraman.

these intrusions, and show that they, unlike other squadrons, could plug the gap. The recce intruders, having completed their missions over the northern cities, would cross the narrow stretch of water and return down the west coast of Britain. The Strangford squadron put forward the case that perhaps sources in Éire were feeding the Luftwaffe with the movements of the patrolling fighters, so it

Fleischmann, took this photograph over the Bristol Channel on 1 April 1941, as a trio from the Luftwaffe base at Tours attacked shipping,

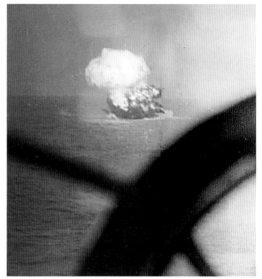

A direct hit is scored on the tanker *San Conrado*.

Damaged by anti-aircraft fire, the crew made a wheels-up landing in County Waterford and used a destructive device which failed to operate. They then used a machine-gun to wreck the instruments.

A view of Heinkel 111 H-5, coded 1G+LH, from the 1st Staffel of KG27 showing its 'Boelke' insignia.

developed a ploy called 'Dublin Sweeps'. This involved a section of Spitfires flying down to the Irish capital, blissfully ignoring neutral airspace, like their quarry. The fighters would then maintain a standing patrol to and fro between Dublin and the Welsh coast at an altitude of 25,000 feet and in complete radio silence. Still the lone Luftwaffe sorties managed to get through – but not always.

The Beaufighters of 125 Squadron caught the next bandit (Ju 88D, code 4T+MH) and sent it crashing in flames into the sea off Hook Head in County Waterford. This encounter was minor compared to a clash between many Spitfires and a lone Ju 88 which took place over central Ireland a few days previously, as another chapter records.

Several of the Irish Sea battles were filmed but while engaged in this work, cameraman George Fleischmann had a bad day. It was 1 April 1941. Fleischmann had been a budding cameraman when he assisted Leni Riesenstal, the famous film director noted for her coverage of the Berlin Olympic Games in 1936 and for propaganda films. He was drafted into the Propaganda-Kompaine (PK), a semi-

Medical aid was sought for Lt Grau whose arm narrowly escaped amputation in hospital.

military organisation made up of journalists, photographers, radio reporters, cameramen and even poets, authors, playwrights, printers and cartoonists. All had been turned into war correspondents and bore the name Kriegsbericher. At first, Luftwaffe pilots were reluctant to

An Air Corps salvage team dismantles the Heinkel which landed at Bonmahon. A mobile workshop is seen behind the nose and an articulated lorry on the right.

take PK men on board on actual missions as they understandably did not want 'spectators' getting in the way. To overcome this, the Reich Propaganda Minister, Josef Goebbels, had them trained as gunners and bomb-aimers.

George Fleischmann was posted to KG (Boelcke) 27 as a lowly Gefreiter, or private first class, but he was soon promoted to Sonderfuhrer, a kind of military civil servant akin to the airborne meteorologists whom the aircrews referred to as 'frogs'. On the above mentioned day, not a very propitious date, Fleischmann was on board Heinkel 111 (coded 1G+LH) engaged in anti-shipping operations when it scored a direct hit on the tanker *San Conrado* and he duly recorded this on film. Anti-aircraft fire from an escorting naval vessel scored some hits on the Heinkel and shortly afterwards it was attacked by fighters from 315 (Polish) Squadron.

The navigator, Lt Grau, another Austrian, was hit in the arm, and with smoke and flames coming from the starboard engine the pilot managed to make a forced landing in County Waterford. The aircraft ran on and collided with an earth bank. Fleischmann was first out of the aircraft and, with hands in the air, he learned from the first arrivals on the scene that he was in Éire. The crew smashed in the remains of the Heinkel's glazed nose and removed Grau who was attended by a local doctor in a nearby farmhouse. The crew then attempted to operate the self-destructive device but it did not function. Grau was transferred to hospital where he managed to persuade the surgeon not to amputate his badly damaged arm – which, in due course, recovered completely. April Fools' Day was not only a bad one for Fleischmann and his comrades, but also for the two other Heinkels which had accompanied his. Heinkel

1G+HL was shot down into the Channel and 1G+FL crashed on the cliffs of Lundy Island while attempting a forced landing caused by engine failure.

During his internment at the Curragh, George Fleischmann was promoted to the rank of lieutenant. At the end of the European war, together with three other Austrians, he was allowed to remain in Ireland on condition that all kept a low profile. The four could have faced treason charges had they been forcibly repatriated to Austria. Fleischmann pursued his undoubted filming skills and produced commercial and official documentaries. He married an Irish girl, wrote a novel based on his Curragh experiences called *Die Gefangenen der Grun Insel* ('Prisoners of the Green Isle') and eventually retired to Canada where he was killed in a simple domestic accident.

Condors

Winston Churchill was moved to excoriate the Focke-Wulf 200 Condor as 'the scourge of the Atlantic', but its activities also extended to the Irish Sea. Designed originally as a 26-seater passenger plane for Deutches Luft Hansa (as the German national airline was then called), it was a large, comfortable, fast, four-engined, all-metal aircraft with several record overseas flights to its credit. By the outbreak of war it had been developed into an operational maritime raider equipped with machine-guns, cannons and bombs – but its basic design was not strong enough for warfare.

When the Luftwaffe had settled into its new bases in occupied France, the Condor group, known as Kampgeswader 40 or just KG40, was detailed to harass Allied convoys and liaise with U-boats in the Atlantic. But the unit

This dramatic photograph, much used for propaganda purposes, shows a Condor wearing Stannkennzeiche (factory markings) SG+KS. Its immediate predecessor was SG+KR which in Luftwaffe service became F8+AH of the 1st Staffel of KG40 under the command of Oberleutnant Paul Gommert. It was damaged by anti-aircraft fire from a merchant vessel and flying on at 850 feet in dense mist it crashed into Cashelane Hill in County Cork. With the exception of one badly injured man all the crew were killed. It is seen below with Luftwaffe codes.

east coast though it was encouraged to sheer off by warning shots from the Dublin AA batteries. This did not deter it from attacking the mail boat *Cambria*, the main Irish/UK sea link, with bombs and machine-guns, but to little effect. Not satisfied with its depredations, it turned inland and flew over Dublin Airport before setting a course for the south-east and its base.

also scoured home waters looking for prey. Initially on these missions the Condors flew west across the Bay of Biscay and then, turning north and skirting the west coast of Ireland, flew on to land at airfields in Trondheim or Stavanger in occupied Norway. A return flight to France would be made a day or two later. On a day in December 1940, KG40 was able to put up three Condors (a not inconsiderable number in view of the type's poor serviceability record) and sank the tanker *Osage* a few miles off the Wicklow coast. The crew, virtually unscathed, were picked up by a collier and landed at Rosslare harbour. Meanwhile, the Condor continued up the

The next day another Condor flew close to Dublin, going south until it passed over Rosslare Bay at zero feet. The crew spotted an Irish vessel which it identified as neutral, and then went on to attack the Irish Lights *Isolda* (which serviced lightships and lighthouses in Éire's waters) because the vessel was flying the British merchant navy's ensign, the 'red duster'. The Condor's bombs found their target and the vessel blew up and went on fire. Six sailors were killed: the Condor was certainly a scourge over the Irish Sea as well as the Atlantic.

The flights by KG40 usually had a dual purpose: a

meteorological expedition combined with a raid on shipping – even when the former was the pre-eminent task, bombs were carried, just in case. A Condor coded F8+OK set off on a morning in August 1940 to check the weather following what was to become a well-worn triangular path from Merignac. It flew along the west coast of Ireland to a point close to Iceland. The Met procedure was standard: at various points the weatherman would call for different altitudes so that he could take pressure readings at various levels. The 'OK' in this Condor's code was inappropriate because something went very wrong and it disappeared. The following day a second Condor flew the same course searching for the first, but neither this nor another on the third day saw anything of what had happened.

Three weeks later the bodies of the captain and one crewman were washed up on the Galway coast; the others are still listed as missing. It looked as if F8+OK had encountered the enemy – but there are no British reports to support this. Perhaps structural failure was the cause? There had been previous inexplicable Condor losses when aircraft were known to have dived straight into the sea because of an inherent weakness in the tailplane. As a result of these incidents, later aircraft were significantly strengthened.

KG40's first major success was the sinking of the liner *Empress of Britain* which had been remustered as a troopship. A Condor set the great liner on fire and radioed U-boat 32 to deliver the *coup de grace*. The Condor in this attack flew safely back across the south-west of Éire to its base – there was nothing to impede it. Once again, the Air Corps' lack of modern fighters provided an 'open skies' scenario to all comers. This liaison between Condor and U-boat was to become a significant factor in the Battle of the Atlantic – hence Churchill's comment.

Even though KG40 could only muster a maximum of six Condors at any one time, it was undoubtedly a useful weapon. By 1943, when a relatively modern Irish fighter squadron had been developed, the Condors had disappeared from Irish skies. But before this, five Condors ended up in Éire or its waters. The first of these resulted in a miraculous escape for its crew – as described in the chapter 'The Magnetic Mountain'. Others who followed were not so lucky.

Men in the LOPs and lighthouses frequently observed homeward-bound Condors with flak and fighter damage. The RAF suggested that a special squadron of Beaufighters, with an endurance of up to 1,500 miles, be deployed to patrol

Condor F8+GL, seen here flying wingtip-to-wingtip with another Condor, went on fire while flying close to Long Island off the Cork coast at Schull. It came down in the sea and the crew scrambled into their dinghy and were making for the island when they were picked up by a trawler, many of which were in the area. The Condor had been badly damaged by gunfire from warships escorting a convoy and its crash was the third one experienced by the 1st Staffel of KG40 on that day.

An Air Corps mobile crane tows a laden 'Queen Mary' out of a soggy situation.

Sean O'Foghlu

The remains of a Condor in County Tipperary after it had force-landed and been destroyed as per the usual German practice.

The skeleton of Condor F8+KH perched on Faha Ridge, Mount Brandon, overlooking the lake in which it could have been immersed.

the Condors' known hunting grounds, but this proved impractical. However, two sizeable RAF raids on the KG40 base at Bordeaux cut down the Condor attacks, at least for a short period. A more long-term solution was the cover provided by Catalinas and Liberators when these came on stream.

German navigators used landmarks, and in particular lighthouses, on the long flight between Bordeaux and the North Channel, where the Atlantic convoys converged. The lookout posts were monitoring aerial activity and occasional U-boat sightings too – information which was rapidly transmitted to British Intelligence – highly un-neutral!

The Atlantic marauders did not have it all their own way and on three days in succession in April 1941 a KG40 aircraft

A Condor crew returns after an Atlantic mission. This crew and their aircraft, F8+CK (in the background), were lost on the last day of 1942.

failed to return to base. On the 18th of that month one was damaged by AA fire from convoy escorts and, as it began to lose height, the pilot made for Ireland. He ditched off Schull, County Cork, and the crew took to their dinghies and were picked up by an Irish vessel.

In the early days of February 1941, a Condor, during a reconnaissance to the west of Ireland, spotted a convoy approaching Britain. Over the next two days Focke-Wulf 200 C-3s, a new mark which KG40 had just received, sank two merchantmen. One of the Condors (F8+AH), nicknamed 'Adolf Hitler' from its suffix, bore on its tail an inscription "England 111" which showed that it had made three land raids. One of the vessels, *Major C.*, damaged this Condor and set it on fire. Escaping, it plunged into a belt of heavy fog covering the west of Ireland and concealing the high mountain ranges of Cork and Kerry. The Condor struck the side of Cashelfeane Hill and disintegrated into a fiery mass. Quickly on the scene was Mary Nugent, a veteran of the Old IRA's female contingent, Cumann na mBan, whose members were experienced at first aid. Despite the danger from exploding ammunition, she dragged away the only survivor from the inferno. He was badly injured with burns and a broken leg, and she expertly tended him. He asked her, "England or Ireland?" and when she answered he responded, "Good, good".

The dead were buried in Bantry Abbey with the fullest military honours which included an army band; amongst the mourners was Herr Henning Thomsen, First Secretary of the German Legation in Dublin. An ordnance detachment destroyed the Condor's bomb load which had scattered and travelled some considerable distance down the mountain slope. The survivor was given intensive medical treatment and convalescence at a seaside resort. For her bravery, Mary Nugent was awarded the Reich Civilian Merit Medal accompanied by a letter from Herr Hitler.

By March 1941 the Condors' losses were accelerating. A Beaufighter from 252 Squadron at Aldergrove shot one down about 70 miles off County Mayo. The next day another Condor was lost and on the third day also, another. On 18 April a convoy damaged yet another aircraft and it ditched near the shore between Calf Island and Goat Island off County Cork. The crew took to their two rubber dinghies and were picked up by an Irish vessel and interned. The Condors still persisted in infringing Irish airspace daily, flying at only a few hundred feet.

The final version of the Condor, a Mk C-6 (coded F8+MR) suffered a fire in one engine which forced its captain to abandon plans to attack an Allied destroyer, and instead plot a course for home. It got hopelessly lost and apparently mistook the Shannon for the Loire river until eventually, low on fuel and with its landing lights ablaze, it came down in an open field in County Tipperary.

What happened was earnestly described by one of two young brothers who were eyewitnesses:

> It was one of those evenings we sometimes get in December. Half-darkness, mild, with no wind, soft and low cloud. The heavy throbbing sound of a plane's engines indicated its size, and a powerful searchlight from its underneath illuminated the countryside. As we watched, it flew in a wide semi-circle over the Silvermines Mountains and for a time disappeared before we saw it again flying over the Shannon. About an hour after we had first seen it there was a heavy dull thud and a bright glow some distance away and we knew that the plane had come down. We ran towards it in a state of high excitement and I remember the Nenagh Fire Brigade passing us out.

When the LDF came up they heard voices crying out; they thought that there were people trapped inside the plane. They soon realised that the shouting came from the crew who had taken cover some distance away and were warning them not to approach the aircraft – but in unintelligible German.

Two LDF men cautiously approached the wreck and, as they looked through a porthole, an explosive device threw them back, the one suffering a severe head injury and the other damage to an arm. The Germans came forward at once and rendered first aid. One LDF man lost an eye and, despite treatment, suffered much pain all his remaining years. The German crew surrendered to the other LDF men and they were marched off, being described as "eight young men, cool and unafraid, still wearing their flying jackets". A multilingual LDF officer explained their situation to them.

Ignoring the LDF, souvenir hunters were soon on the scene picking up batteries, pieces of metal, perspex and empty cartridge cases –a small aluminium door from a gun turret with 'hinges' made of leather still survives. These scavengers were foolhardy because tracer bullets curved and flashed erratically, but the army soon cleared the crowd and sealed off the entire area. The aircrew was to spend Christmas in the Curragh internment camp and were the last airmen to join their comrades there – 'til war's end.

Now came the hard slog for an Air Corps salvage team: it was winter, and Tipperary's famously fertile fields were soaked. Into this morass a team of 17 salvagers, with a crane and three low-loaders, took over and spent a week dismantling and retrieving seven tons of Condor. The low-loaders sank to their axles: spare railway sleepers from a nearby station had to be used to get them out. The owner of the farm was dismayed by the resulting havoc to land and ditches but he was duly recompensed by the government. The onerous salvage work, however, was unpaid, unlike incidents involving the Allies who, it must be said, at least received valuable equipment and often intact aircraft. Air Corps expertise increased with every aircraft type encountered.

In the town of Nenagh near the crash site there was an aluminium manufacturing company which shared redundant Condor material with the Ordnance Corps. The factory made pots and pans from the scrap, while the Army used their share for smoke grenade casings.

Innisvickellaune

The Blaskets stand a few miles off the seaward point of the Dingle peninsula and they were the last bastion of an Irish-speaking community which had virtually disappeared by the beginning of the war. Three autobiographies by islanders have, in translation, become classics, including the chronicles of Peig Sayers whose life story was de rigueur for every pupil in the days when the Irish language was still compulsory in the schools.

Close to the smallest island, Innisvickellaune, on a cold day in November 1940, an ungainly flying boat with the middle engine of its three diesels stuttering and spluttering, alighted on a rough sea. The aircraft was a Blohm und Voss 138-7 (coded 8L+CK) from Kustenfliegergruppe 906, which had

Typical of the coastal areas patrolled by the Air Corps in the war years are the Blasket Islands off the rugged Kerry coast. Here, in November 1940, a Blohm und Voss flying boat came down on the sea. Its crew managed to reach the larger island seen in the background and lived off the land for several days before being apprehended.

Irish Tourist Board

been out searching for Allied shipping in the Atlantic. The Luftwaffe called it 'Der Fliegende Klug' ('The Flying Clog'). The type was prone to engine trouble and this forced the 23-year-old pilot, Willi Krupp, to put down at the first opportunity. He chose the sound between Innisvickellaune and the Great Blasket as it provided some shelter. Having failed to remedy the fault, the crew used a machine-gun in an attempt to destroy the aircraft. They did not succeed and finally took to their two rubber dinghies.

The weather was so bad that nobody on the Kerry mainland, or even on the nearby Great Blasket Island, noticed the plane or heard the shots fired by the crew. They used their pistols to blast hand grips so that they could scale the cliffs of Innisvickellaune. They remained for three days on the uninhabited island, lighting fires to keep themselves

warm and also to attract attention. For food they shot a sheep and some rabbits, supplementing these with condensed milk and flour which they found in an empty house sometimes used by visiting shepherds. Willi Krupp had been a chef and baker before his Luftwaffe service, and he made the best use of the fresh meat.

Weary of their 'Robinson Crusoe' existence, the Germans set out for the Great Blasket in their dinghies, but a very heavy sea was running and they became separated. One dinghy was swept away, but its two occupants made it to the island; the other three joined them shortly afterwards and Lt Konrad Neymeyer was brought to the mainland by fishing boat to report their plight; in the German service, the pilot was not necessarily the 'flugzeug fuhrer' or aircraft captain. A Marine Service vessel, *Fort Rannoch*, collected the others who

This Blohm und Voss BV138A flying boat is being piloted by Willi Krupp who was at the controls of its sister ship (from which this photograph was taken) when it landed near the Blaskets in 1940.

Willi Krupp.

were suffering from exposure. All were held overnight in hospital before being transferred to Cork where they were interrogated by an army intelligence officer.

Konrad Neymeyer spoke good English and had a considerable grasp of Irish affairs: he discussed the strife between England and Ireland over the centuries. Assuming that Anglo-Irish relations were still strained, he hoped that the authorities would connive in getting him and his men back to Germany. He was secretive about his mission, saying rather apologetically: "I trust you completely, but we were warned to give no information whatever lest it might perhaps fall into the hands of our enemies." The Irish officer plied the German with whiskey, but later lamented: "The consumption of seven glasses seemed to have no noticeable effect on him – other than diuretic!" The crew were happy with the two nights spent in Cork, though at one point a note was passed to the Irish officer informing him, "Without beer it is good, with beer it is better", and beer was provided.

The German Legation insisted that Innisvickellaune lay outside the recognised three-mile limit but the Irish authorities relied on the Fisheries Protection Act 1933, which extended the limit in a southerly direction from the Great Blasket. Had the aircraft come down outside the territorial limit, the crew would have had to be released under international law. Germany did not recognise the extension of the limit beyond three miles, as her own attempts to extend her territorial waters had been opposed in the past. But repatriations would have been problematic. There were several eye witnesses to testify that the aircraft had certainly flown over territorial waters north-west of the Great Blasket immediately before it alighted. The Legation employed the same argument later when another Luftwaffe aircraft crashed close to the Cork coast, but again to no avail. The three-mile limit was something of an anachronism; it had been fixed in a previous century when that distance was the maximum range of a cannonball!

The flying boat had to be sunk by *Fort Rannoch* which fired a dozen rounds into it: the sturdy armour-plated aircraft resisted until the last round caused a secondary explosion and down it went. Pieces of its struts can be seen today in use as handrails on steps carved out in the cliff where the two crewmen had scrambled to safety. Konrad Neymeyer, while interned in the Curragh, broke his parole by stowing away on a ship in Dublin, assisted by an Irish soldier, but he was apprehended in London. He found conditions as a prisoner of war there, and later in Canada, far less comfortable than those of an internee at the Curragh.

Forty years later, when the then Taoiseach, CJ Haughey, was visiting Bonn, he met the crew. He had a holiday home built on Innisvickellaune and invited them to spend a fortnight there as his guests. The visit was enjoyed by all and the Germans learned that an engine block from their aircraft was now lying on the seabed and was used as moorings by yachtsmen. During his stay in Ireland, the amiable Willi Krupp became a dedicated Hibernophile and in his will

specified that a Celtic cross be raised over his grave. He passed away in 1997.

Aufklaerungsgruppe

The Luftwaffe called a reconnaissance group 'Aufklaerungsgruppe'. If long-range was involved, then in parenthesis the word 'Fern' was added. Thus, the full title of one squadron based at Buc, a few kilometres from the splendours of Versailles, was Staffel/Afklaerungsgruppe (Fern) 123. (The Germans have a penchant for long words.) The unit operated a regular reconnaissance service which was known as 'Paris Zenit' to reconnoitre the UK and its surrounding seas. A single Junkers Ju 88D or sometimes a Heinkel 111 would carry out the mission.

On a clear August morning in 1941, the mission given to the crew of Ju 88D-2 (coded 4U+HH) was to photograph Bristol and Birmingham. The aircraft made the routine stop at Brest to top up its tanks, but when it arrived over St George's Channel it was jumped by two Hurricanes from

The headquarters of KG123 (F) were at this fine French chateau at Buc. Nearest the camera is the commanding officer, Major Coppor, while in the background is one of his pilots who ended up in the less comfortable quarters of the Curragh internment camp.

A senior officer presents a Ju 88's crew with their combat clasps at Buc. At least two of the four crewmen crashed in Éire.

615 Squadron, based on Anglesey. These had been alerted to the German and were soon airborne at 10,000 feet. One of the pilots, F/O Rene Mouchotte, a noted Free French airman, saw the intruder outlined against a cloud and manoeuvred to get the sun behind him before he dived vertically to attack while his companion protected his back. He almost collided with his target and in a severe climbing turn to avoid it, he blacked out. When he came to, the Junkers had disappeared into cloud – in the Frenchman's own words: "thick black rain cloud".

Mouchotte had perfected the 'vertical approach' to the extent that he had 'shot down' his commanding officer 16 times with his camera gun. The latter was delighted, when Mouchotte returned to base, to discover that the tactic had worked in the Frenchman's first combat and first kill. Initially, Mouchotte and his British companion had no idea that the Ju 88 had gone down and were somewhat despondent on landing. When the news came through that he had scored his first victory, and the first for his base, congratulations poured in from all quarters.

In fact, Mouchotte had done sufficient damage in knocking out an engine to cause the pilot, Lt Ludwig Stockbauer, to seek a landing. He came in very low across the south coast, to Belgooley near the tourist and fishing resort of Kinsale, and began to circle the locality looking for a suitable landing site. A few miles further inland Bob O'Reagan had just completed cutting a field of wheat with his reaper-and-binder and was about to untackle his team of three horses. He remembers:

As the plane came further inland I saw a plume of black smoke from one of the engines; it was coming straight towards me but turned slightly to the left and flew on to where it eventually crashed. It circled and came straight at me again at about 500 feet. We had heard of German crews firing at farm workers in England and I feared this fellow was going to shoot at me and my grand team of horses. But when he turned left again I realised that he was about to crash-land – about two fields away the whole top of the plane was jettisoned and fell to the ground. [When a Junkers Ju 88 was about to crash-land, the drill called for the canopy over the rear section of the cockpit, on which the machine guns were mounted, to be dumped. This ensured that the crew could get clear quickly after landing.]

Two hours after the air combat, the Hurricane pilots landed back at their base and were informed that the Y-Service (Wireless Intelligence Service) had intercepted the final radio transmission from the Junkers. It ran as follows: "Have been attacked by two Spitfires; one engine stopped, the other damaged; intend to land in Ireland; destroying papers on board. We shall return after victory. Heil Hitler!" Mistaking Hurricanes for Spitfires in the heat of the action was understandable. Bob O'Reagan continued:

> Seconds after the canopy was dumped, the plane landed on top of the hill. Taking care to tie my horses to the fence, I made off through the fields to the crash site. As I ran I heard an explosion caused by the crew setting the plane on fire. A few men, members of the LDF, who had been working in the fields were coming up to the aircraft. They had been waved back by the uninjured crew, one of whom was firing his pistol at the plane, which exploded in flames. The senior LDF man then took charge of all pistols and an amiable conversation began – the pilot spoke fine English. More people arrived and nobody thought of arresting anyone – we all acted like old friends. Looking back now I think it was the strangest hour I ever spent. Our people in rural Ireland knew very little about the war and here it was in all its stark reality. The attitude of the hundred or so people who gathered near the wrecked plane that day would make a story in itself for they included pro-German and pro-English, anti-war, and so on . . . In fact one young man was home on leave from the Royal Navy. His ship was one of the fleet that hunted down the German pocket battleship the *Graf Spee*, that was scuttled in 1939.
>
> The Gardai and army field engineers arrived, as well as the Red Cross ambulance from Kinsale, whose crew looked after some minor scratches that the airmen had sustained. The fire had gone out by this time and the Army loaded every bit of the aircraft onto lorries by nightfall. We were sorry that we didn't use the time before the army got there to take away some of the broken parts as souvenirs! All was not lost because we still had the cabin clock which had been blown down to the foot of the hill. The canopy had two machine-guns but the barrels were twisted where they had hit the ground. Parts of the cabin were perfect and one section is still a window in a farmer's chicken house. I have got a nose cone in bullet-proof glass and an aluminium cogwheel. One German officer has been back many times but I never met him; I heard he was looking for a piece of the plane but I fear he never found it.

Bob O'Reagan may have been somewhat mistaken about this officer's quest: in fact he was seeking some personal papers which he had not thrown on the bonfire but had managed to hide nearby. Eventually, on a visit, he retrieved them. This was the full story behind the bald statement issued by the Government Information Bureau which stated: "A German plane crashed near Kinsale, County Cork, last evening. The crew, who were uninjured, have been interned." The strict censorship which prevailed in Éire ensured that the minimum of information was released.

Censorship was not, however, applied to conceal the story of Auf 123's last operation over Ireland. One of the

Fiddlemaker Murt Collins, an ex-member of the Corps of Engineers, on the right, and a fellow musician are playing on two of the violins made from the timber remains of the Belgooley Ju 88. The violins are particularly tuneful.

Kerryman

Landfall Ireland

The wreckage of a Ju 88 in Belgooley, near Kinsale, County Cork.

unit's aircraft was assigned to drop two Irishmen into Éire. The two had been engaged in tomato picking in Jersey when the island was taken over by the Werhmacht. They volunteered to work on Germany's side and eventually found themselves broadcasting propaganda from Berlin – but their efforts had very little effect on the course of the war. On landing they were quickly apprehended and lodged in Arbour Hill military prison.

The running fight

Almost exactly a year after the County Cork crash, another cryptic bulletin from the Government Information

About a year before it was involved in the vigorous battle described in this chapter, the Ju 88 coded D7+DK notched up the 500th mission of its unit, Auf (F) 123. It is shown here on its return when it was garlanded with victor's laurels and had a plaque placed on its nose throughout the celebrations which ensued.

Rolf Holterhoff

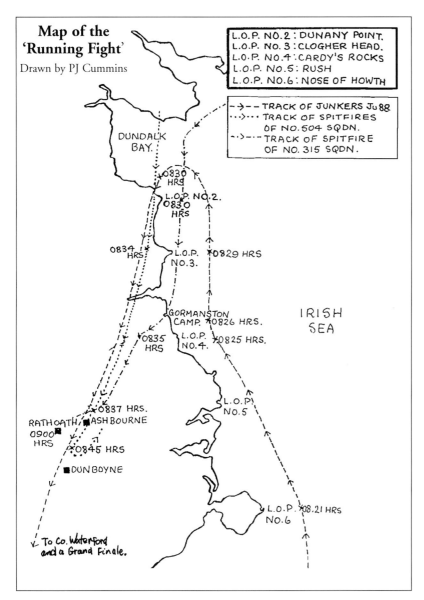

Map of the 'Running Fight'

Drawn by PJ Cummins

L.O.P. NO.2 : DUNANY POINT.
L.O.P. NO.3 : CLOGHER HEAD.
L.O.P. NO.4 : CARDY'S ROCKS
L.O.P. NO.5 : RUSH
L.O.P. NO.6 : NOSE OF HOWTH

- → - TRACK OF JUNKERS Ju 88
- ·›· - TRACK OF SPITFIRES OF NO. 504 SQDN.
- ·›· - TRACK OF SPITFIRE OF NO. 315 SQDN.

DUNDALK BAY.

0830 HRS
L.O.P. NO.2. 0830 HRS
0834 HRS
L.O.P. NO.3. 0829 HRS
GORMANSTON CAMP. 0826 HRS.
L.O.P. NO.4. 0825 HRS.
0835 HRS
0837 HRS.
RATHOATH, ASHBOURNE
0900 HRS
0845 HRS
DUNBOYNE
To Co. Waterford and a Grand Finale.
L.O.P. NO.5
L.O.P. NO.6 08.21 HRS
IRISH SEA

monitored from both shores. Spitfires from 152 Squadron in south Wales took off but failed to make contact. Others were sent off from various bases, including some from the 5th Fighter Squadron USAAF at Derry which was equipped with Spitfires at that time. The first to sight the German aircraft was a pilot from 315 (Polish) 'Deblinski' Squadron. Flying Officer Bolesaw Sawiak opened fire but was given an accurate response by the German rear gunner.

Now a pair of Spitfires from 504 Squadron joined the fray and over County Meath their pilots observed that Sawiak had broken off and was starting to descend. The Pole had been hit twice and, despite his wounds, made a good approach to a forced landing – until his Spitfire hit a rock, the only obstruction in an otherwise clear field, and disintegrated into three sections. Though the local doctor attended to his wounds and had the badly injured man transferred to Dublin's military hospital, Sawiak died.

Meanwhile the German aircraft, with smoke streaming from its starboard engine, continued southwards at virtually zero feet, being harassed by the remaining fighters. These were relieved by a couple of 152 Squadron Spitfires which put the Junkers' starboard engine out of action. The crew made a run for home but the Spitfires continued to pump out lead from their cannons and machine-guns and set fire to the port engine.

The Ju 88's pilot, Oberleutnant Paul Stoermer, described that hectic day:

> We had a short overnight rest at Brest, during which I had been unusually troubled by premonitions of a coming disaster. We took off at break of day because the best time for photography was about two hours after sunrise. We came into the southern entry of the Irish Sea near Cork flying very low, at about 50 metres, so that the

Bureau announced a similar crash in County Waterford. A Ju 88D-1 (coded 4U+KH) again from Auf (F) 123, set out to photograph Belfast docks and any unusual naval activity generated by the Americans who were now moving over to Europe in large numbers. On the previous day the aircrew, with a different pilot, had flown a similar operation, eluding the attentions of 540 Squadron, based near the border, and returning safely to base, guided by its navigator who was a veteran of the Condor Legion in the Spanish Civil War.

Skirting the east coast, the Ju 88 was continuously

Pilots of No 315 (Polish) 'Deblinski' Squadron at their base. On the extreme left is F/O Bolesaw Sawiak who was shot down and killed in County Meath during 'The Running Fight'.

English radar systems would not detect us too soon. Everything went according to plan and we were all in good spirits though on tenterhooks lest we be spotted. Our crew was well used to each other and had a lot of front-line experience – unlike many other crews. Coming into the Irish Sea I climbed to 8,000 metres in good weather with no clouds, which promised us good pictures. We didn't know, of course, that we had been spotted and that fighter squadrons had been alerted.

We were now returning and near Dublin, at a height of about 3,000 metres, the first three Spitfires attacked us and pushed us down, causing smoke to come pouring out of our right engine which I had to switch off. Hunt, a veteran of the Condor Legion in Spain, was making good use of his cannon and later we learned that he had shot

down one attacker. When we had shaken off the Spitfires I had to concentrate because matters looked grim: without the right engine and because of other damage I could hardly hold the aircraft straight. I began to look for a possible landing ground as we flew over the Curragh Camp – little realising that we would be interned there that same evening. The navigator, Brendt, persuaded me to try and get back home; while considering all the possibilities I kept to a southerly course, realising that to cross the Bay of Biscay would be very difficult. Luckily – and I cannot say otherwise – over Waterford two further Spitfires were waiting for us; now there was only one thing to do: go down as fast as possible before they could finish us off. Though we still had ammunition and enough fuel to get us back to base and the damaged starboard engine could be switched on for short periods, this was just enough for an emergency landing. We eventually finished up against a stone wall: the plane was burning fiercely and there was danger of an imminent explosion, so we got out as quickly as possible – the war was over for us!

In the final phase of the fight, ammunition cases spewed down on the people making their way to nine o'clock Mass, ricocheting off roads and walls and causing the faithful to dive for cover. Before the forced landing, the crew jettisoned guns and ammunition boxes and the radioman

The burning Ju 88. The fire had already started in one of its engines during combat.
M Frewer

Condors and Heinkels

tapped out a last transmission to base.

Now on the ground the Germans quickly got clear of their aircraft, which was well ablaze, and took cover behind a farm building. A young man approached and brought them home to his farmhouse where his sister provided a meal while he showed them on a map exactly where they had landed – much to their delight. Thus ended a running battle down the centre of the country in which eight Spitfires had been involved; the incident certainly underlined the determination and durability of their Luftwaffe opponents. Later, the German Minister in Dublin protested about the Spitfires making free with Irish air space – but did not mention the operational nature of the Junkers.

The only fatal casualty of this running fight was the 24-year-old Polish pilot, Bolesaw Sawiak. He had completed one year of a three-year course at the Polish Air Force Academy when Germany invaded his homeland on 1 September 1939. With the rest of his class, he was immediately commissioned and when Poland was overwhelmed he, like many others, found his way via France to the Royal Air Force and further fighting. After his final fatal combat, the Irish army accorded him full military honours as his remains, draped in the red and white of Poland, were repatriated to England. He would not be the only Polish airman to die under Irish skies.

The crew, despite almost seven hours in the air and the stress of combat, were in very good spirits, but disclosed nothing to the Intelligence Officer – only mentioning that they had recently enjoyed a holiday in South Tyrol. The German report of the incident ran as follows:

A total of five reconnaissance aircraft from 1(F)/33 and 1(F)/123 were engaged with two weather reconnaissance aircraft from Wekusta 51 and Wekusta 2 on flights over or near the British Isles. With the exception of a Ju 88 from 1(F)/123 all operated over the eastern and western portions of the English Channel. The Ju 88 which was lost had been engaged on a photographic reconnaissance mission over the Belfast area and according to radio transmissions had force-landed in Ireland after an air battle.

What a battle that had been!

The guns which the aircrew had jettisoned landed in a cornfield but the owner kept quiet as he didn't want his standing crop to be trampled by troops. At harvest time he 'found' them and handed them over to the Gardai. An elderly lady complained that her cottage at Ballyknock

A young Paul Stoermer in pre-war days.

Landfall Ireland

Baldonnel in the summer of 1944 is still a grass aerodrome; its name and that of the country is set out in large letters at the centre of the field.

The Air Corps

had suffered a hole in the porch, a broken window, slates damaged and a hen killed by the Junkers jettisoning equipment and by the explosion of the crash which, incidentally, was all of one mile away. But a neighbour reported that the hole in the roof had been there for three weeks previously and the old lady had to amend her claim for compensation!

Safe havens

Not every warplane looking for a landing place had to contend with rough fields and treacherous mountains. There were many safe havens if aircrews were lucky enough to reach them. Maps issued to USAAF crews carried pilots' notes as follows:

There are 19 airdromes [sic] in Northern Ireland suitable for landing four-engine aircraft. The border between Northern Ireland and Éire is shown on the map. Pilots must not land in Éire except in emergency conditions involving the safety of their aircraft.

[A list of nine aerodromes in Éire followed:]

1. Rineanna: First class field, 4 concrete runways, three of which are over 5,000 feet long; if emergency must be made in Éire, use this field if possible, otherwise use No. 2 or 3 below.

2. Collinstown [which had become Dublin Airport on 1 January 1940]: Triangular field; grass runs of 5,100 and 4,200 feet; should be OK for all types aircraft.

3. Baldonnel: 4 grass runways, longest 3,270 feet; should be OK for all types of aircraft.

The following six airdromes are emergency fields

with no service facilities; depending on the approach and condition of the surface large aircraft have to crash-land with gear up to use these fields.

One of the six about which the notes went on to give details – Rathduff – was supposedly a 'secret aerodrome', built specifically for a 'Doomsday' situation in which RAF aerodromes in the south of England had been overrun by invading Germans in the early days of the war.

At Dublin Airport, war conditions dictated that there was just one commercial service a day, sometimes augmented to a hectic total of two. The field was staked, but a small permanent gap was left to facilitate landings by Aer Lingus and the Air Corps. At night the army placed mobile barriers across these gaps because there was a strong feeling that Collinstown was ideal for an airborne invading force. The aerodrome was also protected by an infantry battalion and anti-aircraft guns.

The first foreign visitor arrived on 28 January 1942 in the shape of an RAF Hawker Hurricane, piloted by an American, Sgt Salvatore Walcott, who had been training with the RAF. He left his base in Cumberland but lost his way and was down to his last 12 gallons of fuel. Perished with the cold, he decided to land. After being refreshed with traditional Irish 'anti-freeze', he departed the next morning.

Over a year later the next straggler, a USAF B-17F Flying Fortress, arrived, this time with plenty of advance warning. Lookout posts all over the country had followed its progress as it wandered aimlessly. At midday it turned towards Collinstown where the ack-ack guns gave a pyrotechnical warning. As the B-17 touched down, infantry advanced and quickly surrounded it.

The aircraft's crew of ten were delivering the aircraft to Europe from the US via North Africa. They had been flying continuously for 16 hours and there appeared to be no love lost between the pilot and the navigator. After rest and relaxation for the crew and fuel for the plane, the travellers continued their journey to Britain.

A couple of months later, an RAF Anson, the first of three, arrived at Collinstown. The excuse of the first crew was 'hopelessly lost'. The second had been practising gunnery against a towed target when the weather closed in and with a few gallons of fuel left, the pilot was about to ditch his plane off the Dublin coast when he spotted the aerodrome, despite the absence of a concrete runway pattern. The Air Corps provided fuel, refreshments and a route forecast, and the plane left later that day. The last of the three was from Air Navigation School in County Down, but when over the Kish lightship in Dublin Bay, the port engine gave trouble and the crew made for Collinstown. Aer Lingus repaired the burst oilpipe which had caused the difficulties and the bar was enriched by the sum of £1 7s 7d. (£1.38). This covered six whiskeys, two rums, four soft drinks and cigarettes.

Another couple of 'Faithful Annies' – the affectionate name for the Anson – did not find such safe havens. As the war was drawing to an end an Anson so managed to embed

During the days at Gormanston when the last Ju 88 was awaiting its new British owners, a Short Stirling, which was a large four-engined British bomber but was now acting as a glider tug, came in. It had a rather cocky American pilot and a British crew and had been lost and low on fuel when it made its way into the small grass airfield. A few days later, with the help of a strong headwind, the pilot managed to get the aircraft out again.

itself in a County Roscommon bog that F/L Rory O'Moore decided to torch it, all useful equipment having been removed. Shortly after the end of the war another Anson on navigational training managed to set down in Limerick docks though it was just two minutes away from Rineanna which was now flourishing as Shannon International Airport. The crew was able to walk ashore through the sticky mud and the tidal Shannon took care of the aircraft before it could be salvaged.

On 12 March 1944 an obsolescent RAF Whitley, with four Canadians and an American on board, came in. It had taken off from a base in England but had drifted off course because of high winds. After seven hours, the crew thought they were over the south of England, but the pilot received a considerable shock when he saw the large 'Éire' sign at the centre of Baldonnel aerodrome. He too thought of ditching at sea, but made for Collinstown, sending out a distress signal. It was Sunday and the airport was closed and given over to grazing sheep. The Army cleared the main runway and a soldier stood out with a sheet acting as a windsock. A green signal was fired and after landing, the tired crew were provided with food, cigarettes and blankets, and duly went to sleep in their aircraft which was hangared for the night. Refuelled, the visitor took off for Northern Ireland the next morning.

One day in December 1941 a British civil DC-3 came in to Collinstown from Bristol and disgorged a party of

The Anson that came down in Limerick docks.

Edgar Heenan

civilians. The Air Corps duty officer quizzed his seniors, but was brusquely told "to mind his own business". The party was whisked by car down to Foynes for a flight to Lisbon. Far from being 'civilians', the travellers were various senior army, navy and Foreign Office personnel and one young WAAF officer. The latter was in fact an important member of the RAF's 'Y Service' whose job it was to monitor Luftwaffe signals. With the end of the Blitz on the UK, 'Y' was turning its attention to the battles in North Africa. The secrecy at Dublin and Foynes cloaked another example of Éire's so-called neutrality.

8 More Heinkels

Left: Navigator Arthur Voight in casual 'civvies'.
Rudolf Hengst

Right: Gerhardt Ristler, the radio operator in Voight's aircraft.
Rudolf Hengst

Oberfeldwebel Voigt: the navigator's story

Arthur Voight was an infantry recruit in the pre-war German army when he was peremptorily ordered to transfer to the Luftwaffe. In the air force he became a navigator and, as war loomed, his training became very intensive indeed. He was then assigned to a squadron whose pilots included Prince Louis Ferdinand von Preussen, a grandson of the Kaiser's! The Prince had been a senior executive in the German national airline, then known as Deutches Luft Hansa. During the 'blitzkrieg' in France, Voight was on a bombing mission over Lille aerodrome when heavy bursts of machine-gun fire from some quarter knocked out his aircraft, causing his injured pilot to make a landing in Belgium just when the Low Countries were being invaded. The wireless operator was killed in the action and Voight suffered a bullet in his jaw. After brief hospitalisation he was soon in a prisoner of war camp where he endured severe conditions before being released by the all-conquering Wehrmacht.

When his injuries had healed and after convalescence at his parents' home, Oberfeldwebel (Company Sergeant) Voight was posted back to his old squadron which was now stationed near Tours. Here he was told that of the six machines which had raided Lille, his was the only one to reach the target, for which he had been awarded the Iron Cross, Second Class. Another glittering prize accompanied the decoration, his long-overdue navigator insignia. On the debit side there were many missing faces in the squadron. Voight gradually got back on operations and after Christmas his Staffel was out raiding British targets off what he described as, "That little-known, but big island on the most western periphery of Europe – Ireland!" Then came the news that his squadron was being transferred to an airfield north of Brest, the big French naval base. Already stationed there were the Condors of KG40 which the newly arrived squadron was to assist in the Battle of the Atlantic. The newcomers were told that the U-boat fleet was being temporarily withdrawn for modifications, so the Luftwaffe

Landfall Ireland

Lieutenant Alfred Heinzl, pilot, and Uffz Gerhardt Ristler, flight engineer, in front of an aircraft similar to the one in which they crashed.

PJ Cummins

had to take up the slack. The He 111s were tasked with attacking convoys at the point where they split up and where ships were most vulnerable. The briefing included a warning about transgressing Éire's three-mile limit and regarding the necessity of destroying aircraft should they be forced to land in that country.

As he lined up for his first anti-shipping mission, Arthur Voight was pleased with himself. He had just made a date with a pretty French girl who was employed in cleaning billets at the base at Morgat. He described the scene himself:

Hauptmann Wissmann came straight over to us and said, "The ship you attacked yesterday could not be found and we assume that she has sunk as she could not possibly be in port in that short time. Well done!" We smiled at each other with satisfaction – this kind of praise goes a long way. Back in the briefing room we got the latest report from the Dornier XVII reconnaissance planes. The Hauptmann told us that more shipping had been observed in the Irish Sea but we would have to go further out to get at them. We looked at the map, some of us started scratching our heads until Heinz Kessel came up with the most logical plan. "We had better keep to the Irish side of this bloody St George's Channel until we have located the ship and after our attack go back swiftly to safer waters." We all agreed with this as there was no point in getting acquainted with RAF Blenheims and Wellingtons, or Sunderlands for that matter. Besides, it was not our job to fight air battles, we were told to sink ships, not planes.

The weatherman preached his short sermon, the radio frequencies were detailed and we slipped once more into our battle dress. The sergeant in charge of our ground crew was waiting for us by our machine, 'Heinrich'. He told us that they had searched for a bullet which had struck the aircraft on the previous day but

found nothing saying, "Are you sure it was a bullet?" I told him, "Look at this hole here, what else could it be?" He responded, "Perhaps a small stone, while starting or landing?" Well, we left it at that, climbed in and rolled to the take off point where we were told that our second machine would be delayed for a short while. We climbed out again to have a smoke and as an ambulance passed by our air engineer, Gerhart said: "I hope we won't need that damn thing!" As I was checking my camera Richard, the pilot, laughed and said: "It's not what you shoot with that camera, it's what you hit with your bombs; you saw a classical example yesterday!" The second Heinkel, 'G for Gustav' rolled up and we were off again to a perfect start in perfect weather. We were soon heading for the mouth of the St George's Channel, our hunting ground. "Watch out for Sunderlands or Spitfires, or anything at all," said Kessel over the intercom. Three pairs of eyes watched the sky and sea in the rear of our Heinkel, two pairs in the front. There was some dark spot on the horizon and coming nearer we knew it was a ship. Flying at very low level and just avoiding the splashes of the waves we dropped our bombs – or so I thought. But our bombs hadn't fallen. I looked very closely at the release mechanism – everything was switched on. What could it be? Then at last I saw a fairly deep mark right beneath the push button: it must have been caused by that stray bullet from yesterday's mission which we could not find. I told the pilot that I could release them with the manual lever and he told me something I wasn't very keen to hear: "We'll make a second attack, using the manual release."

This would be a very great risk because we had lost the advantage of surprise and the ship's crew would be waiting for us with their guns trained on us.

Turning around for the second attack, my heart was beating much faster, sweat was coming out on my forehead; before pulling over the ship for a second time I saw plenty of flashes. The sailors were not idle with their hardware and had plenty of time to aim at us. Richard the radio man cried out that Gerhardt had been badly hit and at the same time the pilot nudged me and pointed to our starboard engine which was spluttering and soon stopped. The port one didn't sound too healthy either and we were hundreds of kilometres away from our base in Brest. I said if we keep to the west by north-west coast we should see land soon, the south-east corner of Ireland as I looked at my map. Richard came over the phone then saying that Gerhardt was dead. I told Richard to give him an injection from the first aid kit, but he had already given him two, and I was reminded of the cryptic remark he had made about the ambulance back at the airfield. However, four of us were still very much alive, though our plane was badly crippled; our eyes were focussed on two things only, nothing else mattered, the behaviour of the port engine and the horizon in the distance – with a good combination of both we could save our lives. The engine was only on three-quarters speed keeping us just barely airborne. Seconds turned into hours; if only we could retain some height, but we were coming nearer and nearer to that dreadful sea. At last I saw it, in brilliant sunshine there was land ahead. I looked at the map, what is the nearest inhabited place of that land? Wexford was the name, most likely, a small town. But we didn't want a town or a village or even one house, we were looking for a place to belly land our crippled machine, if we got that far. Max, the rear gunner, announced that he had made our dinghy ready. Our port engine, our lifeline, was getting worse but that tip of land was now clearly in sight. I fastened my seatbelt as the engine spluttered and died. We were barely ten metres over the water and the pilot gave the controls a last jerk up. The Heinkel responded but seconds later she dropped down like a sack. The beach was right in front of us, we could see a stretch of lovely sand, our very last hope, coming nearer and nearer. We were over land! A fierce crashing noise and heavy bumps resounded and Heinrich came to a halt after a nerve-racking belly landing.

Only a few metres away from the water to our right were sand dunes. The time was 14.30 hours, the date 3rd March 1941, the location Ireland. It was just ten months since I had been shot down in France.

Arthur Voight's 'Heinrich' came down onto a broad stretch of beach: the aircraft floated on and on until the pilot had to force the nose down as he was running out of space. The wireless operator tapped out a last message: "Greetings to all our beloved and to our homeland. I end, comrades." The ventral gondola took the impact on the sand and shingle of Roostoonstown Strand, a strip separating a lake from the ocean.

The lookout post alerted the Gardai at the nearby ferry

The tail of a Heinkel 111, with its tally of 'ships sunk', was all that remained after its crew had destroyed it.

The result of Arthur Voight's destruction of his Heinkel. To the left of the propeller blade two men are keeping watch over the body of the flight engineer.

Nicky Furlong

port of Rosslare, while three men working nearby rushed to render assistance. The gunner was past human aid and his comrades laid his body on a sand dune 100 yards away before activating the usual explosive device which failed. Arthur Voight was also thwarted in his attempts to set the aircraft on fire, so he dismounted the dorsal machine-gun and took it and several ammunition drums up to the dunes where Germans signalled the approaching workmen to take cover. Continuous gunfire failed to have any effect on the self-sealing fuel tanks. The Germans continued to warn a gathering crowd and ignored the Garda request to cease firing. Eventually, after ten minutes and three drums of ammunition, there was an explosion which scattered debris and burning petrol for several hundred yards. The gunner had hit a hung-up bomb which had been intended for the SS *Sinaloer*.

Now, troops advanced and took the Luftwaffe crew into custody and provided them with food and Guinness in a nearby pub. None of them had sampled the 'dunkel brau' before. In the Garda station the local priest and a solicitor, both fluent in German, translated and assisted the crew in reporting by telephone to the German Legation in Dublin.

An Air Corps salvage officer reported:

> The only recognizeable part of the plane is the tail assembly which has broken away just forward of the fin and the outer wings. The complicated Lotfe bomb site, though badly burned, can still be operated. Also, there is a hinged-end portion of a heavy steel tube which projects from the extreme stern of the fuselage. I deduce that [this] is a remotely-controlled launching device from which grenades can be ejected to explode in front of aircraft attacking from the rear.

'Gustav', the other Heinkel, was also badly damaged by fire during its attack on the SS *Sinaloer*. Losing height, it made for Lundy Island in the Bristol Channel estuary, where it managed to land on the only suitable spot of ground. No one was injured and, this time, the demolition charge worked. Years after the war, when the pilot had been repatriated, he contacted the Air Corps in the hope of retrieving the portion of the rudder which bore his victories but alas, this souvenir had vanished – a tailpiece, literally, to the Wexford Heinkel.

And more Heinkels

Early in March 1941 a Sunderland from No 10 Squadron (RAuxAF) was making its stately way looking for survivors off the Kerry coast when it was jumped by a pair of Heinkel 111s from KG27. The lighthouse keepers on the Great Skellig rock had a grandstand view of what ensued, even though there were heavy showers at the time. The Germans got more than they reckoned for from what they had nicknamed 'The Flying Porcupine'. The heavily-armed flying boat shot down one plane into the sea, where it quickly sank, while the other Heinkel was so damaged that it made for home, but only covered 30 miles before ditching.

Shortly after the battle, the Sunderland made a leisurely investigation of the area but it left without signalling to the lighthouse whether it had seen anything. The Irish Lights vessel *Nabro* found wreckage and oil on the sea about three miles south of the Rock, but without sign of life. When the Australians got back to their base in Wales, just one single bullet hole was found in their aircraft.

On the night of 5 May 1941 the Luftwaffe was out in force blitzing the dockyards at Glasgow, Greenock,

Liverpool, Newcastle and Plymouth. There were also a dozen more aircraft engaged in anti-shipping operations against convoys in the adjacent sea areas. Anti-aircraft gunners claimed to have damaged at least two He 111s, V4+GK and V4+DK, attached to a KG40 Staffel. It was the roaring engines of the latter which woke up the residents of a coastal village in County Wexford at about 1.00 am. The disturbance ended with a loud crash out at sea. The Heinkel had come down between the Blackwater lightship, anchored about three-and-a-half miles out, and the shore.

The pilot and one of his crew survived when the aircraft struck the water in a series of bumps: the first impact had shot one of the injured men out of the aircraft; after a couple more bounces the Heinkel began to settle down in the water, taking the other injured airman with it.

The survivors managed to get away in a rubber dinghy which drifted for two hours before beaching on a small strand. To attract attention they fired flares into the night sky which were seen by the local parish priest who was ministering to a sick parishioner. He hastened to the source of the signals where he found the two men who told him of their missing comrades. The priest continued to search but could find no trace of them, so he brought the survivors to a house nearby where they were fed and then took them to his parochial house and put them to bed. At mid-morning the next day he drove the two refreshed flyers to the Garda station where they were held until a military escort arrived.

The army concluded that Liverpool had been the target of the Heinkel but this was a supposition, probably based on misinformation from the airmen; they had probably been on an anti-shipping operation, which was KG40's prime role. The Heinkel pilot, Oberleutnant Walther Hollborn, said that his aircraft had suffered from severe AA fire, which had injured two of the four-man crew and had caused the Heinkel to vibrate violently. He decided that he would rather land in Ireland than in England. He said that he could have come down on land but decided to alight on the sea outside the territorial limit so that he would be treated as a 'shipwrecked mariner' and thus be allowed back to Germany – he may well have been within his rights.

The body of one of the missing pair was washed ashore but there was no trace of the other man who had probably gone down in the wrecked aircraft. Both survivors were interned at the Curragh, but Hollborn availed of the parole system whereby he could study in Dublin. He was a contemporary of the author's at University College, Dublin, where he always described himself as 'Walter Von Hollborn', though none of the records include this aristocratic Prussian cachet.

On a Sunday in the spring of 1942 the keepers of the Bull Rock lighthouse found themselves playing host to an unexpected party of visitors. They had been surprised to see a rubber dinghy bobbing on the water and being paddled by four German airmen. Exhausted, and suffering from cuts and burns, they had endured 24 hours at sea and had to be assisted in landing onto the Rock. On the previous day their Heinkel, He III H-6 (F8+ET), had been out over the Atlantic on a convoy bombing mission, when it was spotted by a Bristol Blenheim (QY-U) of 254 Squadron

Back at base the RAF Sunderland from No 10 Squadron (RAuxAF) which had seen off the two Heinkels and sustained just one bullet hole is attended to by its 'ground' crew.

The local group of the LDF rendered military honours at the burial of Friedrich Schutz whose body was the only one recovered from a crash into the Kenmare River (actually a bay in Kerry) by Heinkel 111H-5 of VIII/KG27 on 23 February 1941. Note the swastika flag and the bottle of holy water.

which pursued them for several minutes. The faster British plane got within range and its gunfire instantly ignited both of the Heinkel's engines, causing the stricken German to ditch immediately about six miles south-west of the Bull Rock. The crew escaped in the dinghy minutes before their plane slid beneath the waves.

The airmen, all NCOs, were unarmed apart from a Verey light pistol. The keepers administered first aid and provided food, clothing and beds for the night. Next day an SOS brought the Irish Lights tender *Nabro* from Berehaven to collect the stranded Germans, who were now in good spirits. The cause of the distress signal was not known until the vessel arrived back with its passengers that evening. The Department of Defence had to fork out £5 14s 7d for the hire of the *Nabro*. In due course the Germans were conveyed to the internment camp at the Curragh where they remained for the duration of the war. Again, wartime censorship prevented any word of this incident appearing in print.

9 The 'Magnetic' Mountain

In August 1940, when the Battle of Britain was at its height, the coastwatchers at the lookout post at Brandon Point on the rough Kerry coast heard a multi-engined aircraft close by but were unable to see it, visibility being down to a few yards The plane was later identified as a Condor of KG40 (coded F8+KH) which had left its base at Abbeville in occupied France on a meteorological mission – but also on the lookout for targets of opportunity at sea. 'Der Nebel' as the fliers christened all meteorologists, was busy taking readings. After three hours, the aircraft climbed to 5,000 feet and turned for home. Though the crew thought they were over the ocean, a break in the overcast showed that they were, in fact, over the 'Kingdom of Kerry'. The radio operator could not get a bearing so the pilot decided to head out to sea. Seconds later, flying straight and almost level, he unwittingly flew so low over the lower slopes of Mount Brandon, which rises to over 3,000 feet (953 metres) and dominates the area, that the crew could feel the shaking caused by 'ground effect'. With great good luck, the Condor was flying in a slightly nose-up posture, parallel to the rising ground, before it struck a large rock and disintegrated.

Mount Brandon in County Kerry where one German and three British aircraft came to grief. Tests were carried out to determine if the mountains had magnetic properties which upset compasses – however, bad weather and pilot error were the most likely causes of these crashes. In the foreground is a curragh, the typical canvas-covered fishing boat used on the south and west coasts.

Irish Tourist Board

Left: Lamb's Head Lookout Post No 33, near Mount Brandon, as it looks today.

Below left: The tail assembly of the Condor which crashed on Mount Brandon. The tail was 'the weakest link' in the Condor's structure.

Below right: The wreckage of Condor F8+KH perched on Faha Ridge, Mount Brandon, overlooking the lake in which it could have been immersed. The aircraft came to rest close to the edge of a sheer cliff.

All Cdt O Quinn

Some of the lucky Condor crew returned to Mount Brandon in 1988. Pictured here (left to right) are Kurt Mollenhauer; the daughter of Mrs O'Connor who greeted the strangers with a jug of milk; Kurt's wife; and Kurt Kyck who lives in Ireland.

PJ Cummins

In what was undoubtedly a miraculous escape, only two of the crew suffered minor injuries. The captain, Oberleutnant Kurt Mollenhauer, had suffered a broken ankle and another flier had an injured back; both were soon patched up in the County Hospital. Others whose planes impacted on this treacherous mountain would not to be so lucky. The Germans were soon counteracting the shock and celebrating their escape with bottles of brandy. Soon, over 50 people had gathered at the scene and the aircrew shared out their chocolate with the first comers. The lucky survivors were helped down to the tiny village of Faha, which was Irish-speaking, republican, and generally hostile to the government and – let it be said – sympathetic towards the Germans. The full weight of strict press censorship had not yet been applied and the *Kerry Champion* was able to report that the crew "appeared to be men of fine moral and physical fibre, and were picturesque figures in their brown knee-top boots, leather jackets, and curiously tilted soft caps which they wore in place of flying helmets".

It was an amiable gathering: non-warlike subjects such as current Gaelic football matches, about which the Germans were quite knowledgeable, were discussed, despite language problems. When the injured were sufficiently recovered, all

A memorial plaque to all airmen who perished on Mount Brandon is to be seen in the town at the base of the mountain.

PJ Cummins

six were transferred to No 2 Internment Camp at the Curragh – where Mollenhauer was to become a constant thorn in the side of his guardians.

The war raged on, but Mount Brandon was not in the news again until 1943 when a series of dreadful crashes involving Allied aircraft created speculation that something in the mountain was upsetting aircraft compasses and luring planes to destruction. At the request of Britain, the Air Corps carried out a series of tests which proved to be totally negative – the dreadful weather which often engulfed Kerry was the culprit in all cases. In July, a civilianised BOAC Sunderland Mk III flying boat on a routine service from Lisbon to the UK via Foynes was shattered on the mountain. It had been making its approach to Foynes flying-boat base on the Shannon when contrary winds and bad visibility drove it into the mountain with the loss of the first pilot and nine passengers. However, six of the crew and nine other passengers escaped.

The aircraft's cargo included sackfuls of mail, the first correspondence from those unfortunate enough to be taken prisoner by the Japanese. Some of the survivors were RAF aircrew who had come down in neutral Portugal and had been briefly interned there. One of these described his particular experience:

The tail section of the BOAC Sunderland which crashed on Mount Brandon still shows the last letters of its registration.

Landfall Ireland

The 'Sunderland episode' started for me when I was flying a Mark XI PR Spitfire out to the Middle East via Gibraltar and was forced to crash-land in Portugal. I was interned with a few other aircrew, and after about three weeks in the luxury Elvas Hotel in Lisbon, during which we had the freedom of the city, we were allowed to go home. We boarded the BOAC Sunderland 'ES' which was on a routine flight from the Middle East to England with servicemen and civilians.

We took off from Lisbon at night, and I slept peacefully until we apparently arrived over Foynes in cloud in the early dawn and flew back out to sea to reduce height before returning below cloud level. My seat was immediately below the wing, and I climbed up on it to have a good look out the porthole but I could not even see the wing-tip float directly out from me – only solid mist rushing by. I was sure we were diving off height, an uncomfortable feeling when another pilot is doing it.

Suddenly there was a break in the mist and I not only saw the wing float but also, immediately below, the dark outline of land sloping the way we were going. At the same time the pilot put all four engines through the throttles gate, and I thought 'This is it!' When I came to, I was lying half embedded in mud and slush with the aircraft burning around me in the misty murk and I slithered and crawled away from the wreckage.

The Spitfire pilot's theory was that the Sunderland was off course and instead of keeping over the sea, its pilot cut across a headland and hit the mountain. He also deduced that the pilot had seen the land through the mist but had only time to engage full emergency boost (going through the throttle gate) and get the nose up before the Sunderland crashed on the mountain and burst open like an egg.

Four weeks afterwards, Sunderland DD848 of 201 Squadron, on an anti-submarine patrol from Castle Archdale, struck a marshy patch on Mount Brandon and was completely burnt out, leaving eight of its 11 crew dead. The height and nature of the site prevented the Air Corps from salvaging anything. One of the three survivors came from Limerick and he was allowed to visit his home before being shipped across the border with his two companions.

There were no survivors from the all-Polish crew of a Wellington Mk XIV of 304 Squadron, based in south Wales. Again the iniquitous Kerry weather caused their downfall. The pilot, F/S Klemens Adomowicz, at the ripe old age of 32, was the eldest member of the crew of six of 'HF 208'. Sergeant Hirsz Pawel Kuflik, the navigator, was Jewish. Though born in Cologne, he was brought up in Poland from where, like the others, he fled from the German invasion, later joining the Polish forces in Britain.

The theory of some strange magnetic force emanating from Mount Brandon being discounted, the verdict was that the cause of all these disasters in one spot was that any aircraft was vulnerable in winter conditions, particularly when operating over mountainous terrain. Navigation was still dependent on relatively primitive aids and operational requirements meant that air crew were constantly flying 'on the edge'.

From the site of all these tragedies and, of course, the one lucky escape, one can overlook a most picturesque scene, and it is hard to visualise the terrible things that occurred there during the war years. The view takes in the surfing beaches to the west of the mountain and, down below, the hordes of tourists visiting the ancient oratories and sites of mystical rocks engraved with an early form of Celtic writing – altogether a different magnetic lure in more peaceful times.

10 Junkers

Early in March 1942 a Junkers 88 A-4, coded CN+DU, from Wekusta 2 (a unit dedicated to providing information to the Luftwaffe High Command) was on a Met flight south-west of Ireland. There was a heavy mist and fog with visibility down to five yards. The coastwatchers at LOP No 29 heard the aircraft and minutes later it crashed into the southern slopes of Mount Gabriel in County Cork, killing all the crew. The first to find the wreck was a hunter who, despite the weather, was out after snipe. Heedless of the

The Bull Rock Lighthouse as viewed from a Ju 88.

Rolf Holterhoff

Rolf Holterhoff, a German businessman intrigued by the crash of the Wekusta 2 Ju 88, commissioned this painting and line drawing of the aircraft a few seconds before its disastrous crash. The line drawing (right) portrays the crew. Both illustrations, by the well-known British aviation artist Michael Turner, are reproduced here by courtesy of Herr Holterhoff.

dead, he retrieved a code book which he tucked into his game bag and headed for home. On the way he met two Gardai who were aware of a crash and the explosion which immediately followed it. The hunter merely indicated the general direction where he thought the crash had occurred, but did not mention that he had been to the site and had retrieved the code book. Eventually the Gardai, the army, and local residents arrived on the scene. The going was so heavy that it took all of three hours to get there.

The family of the hunter had, in the earlier part of the century, owned a thriving hotel in Glengarriff on Bantry Bay, the huge deep-water anchorage capable of accommodating the entire Royal Navy. The place was patronised by officers of the fleet and its proprietors were naturally immersed in naval matters – that is until 1922 when the Anglo-Irish Treaty caused the navy to depart. Though they were now in reduced circumstances due to the absence of their former guests, the family kept up its naval contacts. The hunter, now a jobbing jack of all trades, reported regularly to Naval Intelligence on anything he thought might be of interest. He sent on the code book to his contacts who were delighted because it contained a new cipher system which had just been issued to the Luftwaffe. The 'spy' was awarded £25 – a considerable bonus in those times – which he quickly converted into liquid refreshment. This was not the only occurrence in which information was passed on, bypassing official Irish channels.

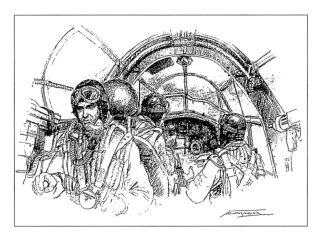

County Cork was also the scene of a later fatal crash of another Ju 88 from Wekusta 2, observed insofar as it could be, by the men of Lookout Post No 32. The post had been moved from Dursey Island off the Bere peninsula to the mainland where Bellinacarriga Hill would provide a higher viewpoint. The coast-watchers were well used to Ju 88s coming in each morning from their base at Nantes. From the beginning of June 1943 to the end of July, aircraft had appeared in the early hours to take weather readings. At this time a young pilot, Hans Auschner, was at the controls with Bernard Nord, ten years his senior and considered to be 'an old man' by the rest of the crew, whose Iron Cross denoted that he had taken part in at least 30 sorties. From late June onwards they used D7+DK, adorned with the unit's weathervane symbol on its nose. Its crew relied on transmissions from Rennes, Bordeaux, Droitwich, and Cork for navigation 'fixes'. These fixes could be up to 40 kilometres adrift in a flight of eight hours. For a reliable final pinpoint the crew would rely on the Bull Lighthouse to indicate the Irish coast and would normally approach it just above sea level.

But one July morning was far from normal and the coast-watchers at the newly-sited Dursey Island LOP did not expect to see the Met plane when the fog was so thick that one could not see one's hand in front of one's face. Nevertheless, they suddenly heard the roar of engines which increased as the pilot opened the throttles wide. At this moment, Michael Murphy, one of the Saygulls, remembers thinking, "Be careful boys, beware of the fog", but at the same instant the starboard wing struck the top of Bellinacarriga Hill. The pilot had been on the point of clearing the hill which he had just spotted through the impenetrable fog. He was seconds too late in taking evasive action. His aircraft plunged into the south side of hill, flipped over the top and tumbled down the other side to a field just 100 yards below the lookout post.

The three coastwatchers gasped with horror as the plane burst into flames. The senior man remembers, after 50 years:

> The plane slewed to the right, striking the ground and breaking off the left wing, at the same time throwing all the crew out of the plane in the direction of its flight. The fuselage tumbled down the other side of the hillside, where it came to rest about 20 feet to the side of the four dead fliers, being stopped by a small stone outcrop. Myself

> and my two comrades found the four Germans about a hundred yards down the slope, lying closely together near the burning fuselage. I will never forget the four good-looking men in blue uniforms with yellow neckties and clean white shirts. The poor lads were still strapped in their seats with their parachutes still in position. Three of them were tall, very young with fair hair, and one was a little man with dark hair and dark complexion – he was Bernard Nord, the meteorologist. In the seconds it took us to reach the wreck it was almost completely burnt out but bullets were exploding in all directions.

On impact, one of the leather-encased petrol tanks broke off and hurtled further down the hill in flames, finally stopping only feet away from a house – a frightening experience for the two brothers inside, but providentially they escaped injury and the flaming tank caused only minor harm. It finally stopped directly under a washing line, destroying Sunday-best clothes and also scorching the posterior of a hapless pig. The terrified animal immediately leaped out of its enclosure, dashed down the hill, and was understandably reluctant to return to its abode until several days later.

A Luftwaffe enquiry concluded that the search for the usual lighthouse reference point was a hopeless undertaking in the prevailing conditions. The verdict was that the crash had been caused by the pilot's fatal error in not remembering the basic rule of relating air speed to range to climbing rate. It was reckoned that from sighting the hill, there remained just 12 seconds for reaction and climbing away, and that a steep left turn would probably have been a better bet than attempting to climb over the ridge.

Many years later, on another misty day, the unfortunate crew were remembered when Col Werner Geissinger of the German air force unveiled a wall plaque on the building on Bellinacarriga which had so narrowly escaped damage. A similar reminder stands beside the tourist car park at Dursey Sound. The two Jumo engines retrieved from the crash were put to good use: one lies in Gairnish harbour as a mooring for boats, while the other was 'liberated' at the time by a travelling roadshow to serve as a generator for its fairground amusements. Swords were turned into ploughshares.

On St Stephen's Day (known in the UK as Boxing Day) 1941, a Ju 88 A–5 of Wekusta 2 (coded DE+DS) was seen to be firing off red and white flares. It was not a belated

A brand new Junkers 88 which made a soft landing in a bog in County Kerry could not be retrieved because of the inaccessible site.

For a change it was the Air Corps which dissolved it in flames – this was a case not of swords being turned into ploughshares, but rather into a pile of white powder!

Christmas display because the Junkers subsequently came inland for an emergency touchdown. The pilot could not have picked a better spot on which to make a sweet landing. This was Caolboig, a vast bog close to Waterville in County Kerry.

Coming in over Ballinaskelligs Bay, it entered the Inny Valley and circled a couple of times while dumping machine-guns overboard. The pilot then made a featherlight landing. In the words of the local parish priest, Father Padraig Sugrue: "At that time a lot of people were pro-German and so the airmen were treated as friends. They had a cup of tea in someone's house before the Gardai and the LDF arrived from Waterville." The crew attempted to destroy their aircraft, as was usual for German airmen, but soon desisted from this.

They reported that their engines' electric pumps were malfunctioning and that fuel was low but an Air Corps engineer and his crew could find nothing wrong with these systems. Indeed, the Irishmen were given to wondering if the crew had had enough of the war. In his own words, the officer observed:

> If I could have got the aircraft back to Baldonnel, one day's work would have made it fit to fly again. The snag was that the nearest bohreen [small road or track] was four miles away and even to carry a few spanners onto

A Heinkel model carved by an interned Ju 88 pilot and presented to one of the people who greeted him following his forced landing in Kerry.

the plane was an operation. If helicopters had been around it could have been lifted out without any bother. The O/C told me to set fire to the lot so the lads splashed fuel on the aircraft's tanks and then got a load of dry grass and heather and strewed it around, stuffing some of it into the mouths of the tanks. When they did this there was a rush of white vapour out of the tanks and everybody ran for their lives as we thought the stuff was going up by itself. I reckon it must have been some reaction between the diesel and the moisture in the heather. In any event we had to set it alight and then sit around all day watching a perfect aircraft reduced to the usual white powder.

Their homely treatment in Kerry was acknowledged later when the aircraft's German commander carved a model aircraft at the Curragh (though strangely of an He 111 rather than his Ju 88) and presented it to an aunt of Father Sugrue's. The priest included the model, together with other artifacts, in a small display collection which was viewed by one of the crew members who returned 30 years after the war.

In Ireland there was no shortage of 'composers' to commemorate such incidents in song and verse. The composer of the following doggerel began his 'epic poem' with a stirring description of the diversions – hunting, coursing, mummers – traditional to the feast of St Stephen

before telling of the day "when a great German plane came in over the sea":

It threw out four flares by Cuan Trae
Saying we're coming to land, in the air we can't stay
In fight with the British far out in the west
We got two wounded, though two of their planes we did best.
It was going half lopsided for something was wrong
Tho' the roar of the engines was powerful and strong
Every eye in the parish was lifted to see
The great German plane which came in o'er the sea.

And of the landing:

Four fine German striplings leaped out on the land
Their documents they burnt with flares near at hand
Five shots at the engines they fired point blank
But they never caught fire for bone dry was each tank
Iron Crosses they wore were from Hitler's own hand
For great deeds of daring o'er sea and o'er land
In the Battle of Narvik they shot down twenty-three
And sent them all blazing down into the sea.

As the priest mentioned above, 'The Kingdom of Kerry' had its own agenda for international affairs – and much else besides!

The last German

By 1 May 1945 the First Fighter Squadron of the Air Corps, now fully equipped with Hurricanes, had transferred to Gormanston Aerodrome in County Meath from Rineanna, which under its new title Shannon International Airport was becoming increasingly busy with civil traffic. A few days later the Third Reich surrendered and when this news was received at Grove airfield in Denmark the personnel of 1/NJG 3 (1 Staffel/Nachtjagdgeschwader 3) were in a confused state and apprehensive about their fate.

Before the news was received, about 40 Ju 88G-6cs were fully tanked up, with crews ready for operations. At least three aircrews decided to fly to other countries to escape from possible attacks by the Danish resistance. Initially these three crews intended to fly to Prague but were unable to do so due to unfavourable weather conditions over Czechoslovakia and

The Ju 88G-6c nightfighter which sought sanctuary at Gormanston Aerodrome a couple of days before the German surrender on 7 May 1945.

an uprising there against German forces. One of the crews then decided to fly to neutral Ireland because the weather forecast there was favourable. The pilot, Oberfeldwebel Gieseke, whose wife lived in South Africa, also thought it would be easier to contact her from Ireland. In the early hours this Ju 88G-6c took off in a westerly direction over the North Sea, crossed the English coast near Middlesbrough and flew across England where they mistakenly almost landed at Liverpool aerodrome. Correcting this error at the last minute, they nevertheless fired a few bursts at ground personnel in a final act of defiance.

Shortly afterwards, and making use of state-of-the-art radar, they let down through a 400-foot ceiling in pouring rain and made an excellent landing on the 900-foot strip of wet grass. They were quickly accosted by an armed vehicle. The aircraft still had sufficient fuel for three hours but this was drained off, leaving enough to run up the engines daily – just in case of a change of heart by its three crew who were taken into military custody and interrogated.

The Air Corps showed great interest in this nightfighter. It was the most potent aircraft of its type in service anywhere, but a shortage of experienced crews and fuel had limited its effectiveness in curtailing the RAF's onslaught on Germany. Its equipment included interception radar installed in the nose with 'stag's antlers' antennae outside; it

'The Last German': at Gormanston an Air Corps painter obliterates the Luftwaffe markings with a British roundel.

had the latest homing radars, secure radio communication systems, landing and navigation aids. A pair of its seven guns were called 'Schrage Music' ('Jazz Music') which could blast a target out of the air when the nightfighter positioned itself in a bomber's blind spot.

A press release issued by the Government Information Bureau stated that "three German airmen landed at Gormanston, County Meath, on Saturday morning. They had flown from Aalborg in Denmark." This bald statement did not prevent British newspapers from speculating that the aircraft had been carrying high-ranking German officers fleeing from Europe. An attempt to put down a question in

Irish army officers with the crew of the Luftwaffe nightfighter which landed at Gormanston in the last days of the war.

The Gormanston Ju 88 had landed four days before this photograph of its erstwhile base at Grove in Denmark was taken. It shows the remaining Ju 88G-6Cs of 1/NG3 parked and, with Teutonic thoroughness, their cockpits covered and propellers removed – even though the war has been lost!

F/L Miller, a Canadian, the pilot of this all blue PRU Spitfire Mk XIX force landed at Spanish Point, Co Clare. This aircraft was a pre-production Spitfire Mk XIX which had been used at Farnborough a year earlier for high speed dive tests. During one test, propeller and engine reduction gear were torn away at almost Mach 1, but the pilot succeeded in gliding back and landing, as shown below.

M O'Sullivan

the House of Commons on whether 'Hitler or other German renegades' were on the aircraft was abandoned.

The RAF was keen to acquiring the Ju 88G-6c to examine its highly sophisticated equipment, and the British Air Attaché opened negotiations which resulted in the Irish government allowing it to be flown to the UK. On 1 June 1945 an Airspeed Oxford I (PH 191), carrying a party of British military personnel with Lt Cdr Eric Brown, a Fleet Air Arm test pilot, in charge, arrived in Ireland. Brown interrogated the Luftwaffe crew about the type of engine, flying characteristics and other details. From their stories it appeared that Oberfeldwebel Gieseke had been born in Berlin and qualified as an airline pilot, but just before the

outbreak of the Second World War he had emigrated to South Africa and married a local girl. The couple had been on a visit to Berlin when war was declared and Gieseke had been called up for military service. He sent his wife back to South Africa and enlisted in the Luftwaffe. His two comrades had equally interesting stories of their wartime adventures.

Lieutenant Commander Brown's team had turned up in uniform, ignoring the Irish request that they should wear civilian attire. Luckily, they had brought 'civvies' with them because, even in the flush of victory, they found that not every regulation was going to be overlooked. Changed, and suitably chastened, they were treated to traditional hospitality. Meanwhile, the black crosses of the Luftwaffe were painted out and RAF roundels were applied to the aircraft. Brown, who was fluent in German, had been briefed by the Luftwaffe crew regarding the aircraft's performance details. Nevertheless, getting out again from Gormanston in this high performance aircraft was 'dicey', even for this experienced pilot:

As the aircraft was light on fuel, I did not experience much trouble in getting it off the grass strip, although it was decidedly tight and I had to run the engines up to almost full power on the brakes, comforted by the thought that if there was no runway I still had lots of power.

The nightfighter, which picked up an escort of Spitfires, then proceeded to the Royal Aircraft Establishment at Farnborough from whence it was sent on to the Central Fighter Establishment at Tangmere. The mention of Spitfires brings to mind that the last wartime RAF aircraft to arrive preceding the last German by a few days, was one such which belly-landed near Spanish Point in Clare. When new it had had a very close shave at the Royal Aircraft Establishment at Farnborough. This aircraft was a pre-production Mk XIX which was being used for high-speed dive tests from 40,000 to 27,000 feet. During one test its propeller and engine reduction gear were torn away at almost Mach 1 – the speed of sound. The pilot succeeded in gliding back and landing, as shown in the smaller photograph overleaf.

One year later the Spitfire was back in squadron service with a Photographic Reconnaissance Unit whose aircraft were sprayed all over in an azure blue which helped to conceal their presence. Speed was their sole weapon; they carried no guns. Flight Lieutenant Miller, a Canadian, made a good forced landing on a beautiful still day. He had just enough fuel to make a second pass at the field he had chosen. He told the authorities that he had been on a training flight but admitted to this writer that he had been actually photographing flying bomb sites, despite the imminent German surrender. The local Garda Superintendent (the author's father) was not happy to see a bright red light on the cockpit panel, though it merely indicated that the undercarriage had not been lowered. Nevertheless, the pilot was asked to disconnect some wires until the offending light went out! Miller's only injury was a bump on the forehead to which some iodine was applied by the local doctor whose son in later years also graduated as a doctor but forsook this calling to become President of Ireland!

11 'The Travellin' Trollop' and other Yanks

On the misty morning of 10 July 1943, the author listened to the sound of a multi-engined aircraft circling over Lahinch, a small resort in the west of Ireland. (Elsewhere, more major events were in train, such as the Allied invasion of Sicily. The morning papers didn't carry this story but tucked inside was a government advertisement for 30 potential sergeant-pilots for the Irish Air Corps, which showed that the country was still hoping to improve its air defence.) A USAAF B-24 was soon clearly to be seen as it descended lower and lower in circles of four or so miles around the west Clare coastline. Shortly, the engine note increased and was followed by a resounding bump as 'The Travellin' Trollop' hit the beach – the lady had finished her travels.

The B-24 Liberator had taken off from Dow Field in Maine to position at Gander for its transatlantic journey to Prestwick. As often happened, the weather throughout its journey turned out to be worse than Gander had forecast, with an overcast sky causing the two pilots to fly on

The crew of the 'Travellin' Trollop'.

instruments for the first phase of the journey. At 'the point of no return', fuel started to leak and this was soon compounded by the radio and the radio compass packing up. The pilots took their plane up to 16,000 feet to allow the navigator to get a fix from the stars which showed that they were over 100 miles south of their proper course, so corrections were made.

Nevertheless, shortly afterwards and over the south of Ireland they were still off their proper track. More corrections brought the Liberator, still in cloud and unbeknown to the crew, over the concrete runways at Rineanna. Eventually the B-24, its undercarriage retracted, touched down on

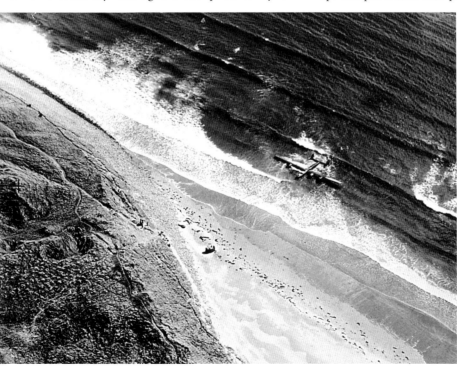

The tide begins to cover 'The Travellin' Trollop'.

Lahinch beach, its right wing-tip digging into the soft sand and spinning it through a right angle before coming to a stop. One propeller tip was ripped off and struck the fuselage just by the captain's seat – he was lucky to escape with only a bruised leg (see cover painting).

A large number of regular army and LDF troops converged on the plane almost before it had stopped shuddering. Unsure of their welcome, the crew, who had evacuated the aircraft, returned to their well-armed plane until they had established that their reception was friendly. The second pilot's first words were, "Geeze, this place is lousy with troops!" and indeed it was, because units of the 23rd Infantry Battalion and the 49th (Limerick) LDF Battalion were quartered less than a mile away at a training camp. Here the airmen were given a good army breakfast and, unusually, allowed to sleep before being questioned. Subsequently they were lavishly entertained in the various bars and hotels of the seaside resort. One gunner, whose ancestors came from County Clare, was so enamoured of his welcome that he ensconced himself in a lavatory and could not be persuaded to come out and rejoin the war! In the end, brute force applied by his comrades persuaded him to proceed with them to the border where they were handed over late in the day after a hectic 72 hours.

A couple of days later an Air Corps salvage team, its work interrupted by the tides, managed to retrieve the engines, guns and various instruments which were also dispatched northwards. However, not everything was returned: despite the presence of an armed guard, several cylinders of scarce oxygen somehow ended up with local medical and engineering practitioners. This was accomplished by native ingenuity when the salvage team and soldiers had to retreat from the incoming tides which regularly semi-submerged the crashed aircraft. On a night-time tide, a curragh (the traditional fishing boat in the west of Ireland) glided towards the tail end of the aircraft, the rowers gained access and spirited the oxygen away, together with the special tools provided by the USAAF. The tools were quickly retrieved by the local police on a 'no names, no pack drill' basis.

When the tide was out, and the salvage team was beavering away, the bored soldiers engaged in bayonet practice against the lady emblazoned on the nose. She was not by any means one of the curvaceous beauties typical of most USAAF 'nose art' but, in keeping with the American tradition, she was topless! There was some talk that the local

priest felt that she should be provided with a painted-on bra! The author had observed the salvage operations with interest, and saw a party of nuns who had a holiday home nearby merrily marching down the long beach, veils a-blowing in the wind, to examine the wreck. As the nose art came into their focus, it was amusing to see their footprints make a wide track in the sand to avoid the brazen hussy!

Chet Miller, who was an assistant radio operator and waist gunner on the B-24, describes what happened afterwards:

> For the fortnight following our brief visit to Lahinch we had a Cook's tour of the UK, taking in all the historical, scenic and picturesque places. That was the start of my sixteen months based in England and North Africa. In that time we made eight successful missions to Norway, Denmark, France, Germany, Holland and Austria. By successful, I mean we managed to get back to base! Several of these missions were real bastards and we got double credit, bringing our score up to 35 missions! All the crew survived the war and by 1944 we returned to the States.

Commandant Jack Ryan, aeronautical engineer and leader of many an Air Corps salvage team.

A memorial to the Skytrain which came down on 17 December 1943.

Vincent McMahon

The crew of 'The Travellin' Trollop' turned out to be more durable in a new B-24 than in their original aircraft. They joined the 389th Bombardment Group of the 8th Air Force and survived at a time when this daylight-bombing force was suffering its worst casualties. In fact the Lahinch fliers were well above average and became a 'lead crew' heading their Group on many missions. Home in the United States at war's end, some stayed on in the post-war USAF – the navigator went on to fly a further 55 missions during the Korean War. He and two other crewmen returned to Lahinch to unveil a plaque commemorating their escape exactly 50 years after the event.

For another Irish village the landing became 'the day the plane came down' and is still talked about.

Dakotas

With the exception of one, all examples of that wonderful wartime workhorse, the DC-3 (or Dakota, C-47, Skytrooper, Skytrain – it answered to several names depending on its duties) which came in, departed under their own power. Engineer Officer Jack Ryan describes how in May 1944, a USAAF Skytrooper came down on Leopardstown Racecourse, chock-full of transport pilots and paratroop dispatchers who were being returned to the UK from the Italian theatre in readiness for D-Day (four weeks later as it transpired):

Our task was to get the Dak out again without having to dismantle it. Fortunately the wind was from the north-east, which gave the longest run, so we hauled the plane back until its tail was virtually against the grandstand – white railings were uprooted by the mile. When I had got an assurance from our Met men that this favourable wind would last for another 24 hours, a company of eager-beaver field engineers got to work. They filled and levelled the ground and felled half a dozen trees with explosives, working till three in the morning, close to a row of houses. Luckily I was off the site during the operation because great panic was caused amongst the residents, with aged grannies and babes in arms being dragged out all over the place, and the poor old field engineer officer being nearly lynched.

' When we arrived back next morning we saw a good semblance of a runway, although it was still 400 yards or so shorter than Aer Lingus' pilots would have accepted. We had lorried all the spare Americans and their uniforms (there seemed to be about nine fancy fighting suits per man) over to Baldonnel so there were just the two pilots aboard at take-off. From the figures they had given me I thought it would be a fairly close thing as they revved up against the brakes, nearly blowing down the grandstand. Letting go, they pushed the throttles through the 'booster gate' and just shot straight into the air. By the time they passed over where the trees had stood, the aircraft was a thousand feet up – what a waste of a stand

Foynes flying-boat base seen
from an altitude of 2,500 feet
in the summer of 1943.

of timber. The Dakota dropped into Baldonnel where its passengers and gear were reloaded, and we threw a party for all: we gathered that most of them didn't expect to live long – here's hoping they were disappointed.

The highest point in Ireland is Carrantohill in Kerry's mountain range known as McGillicuddy's Reeks. It was into this range that a Skytrain flew blindly a few days before Christmas 1943. Unknown to anyone, the five dead crew lay high on the mountain for seven weeks. This belated grim discovery was not made until an off duty infantryman spotted what appeared to be an aircraft's tail wheel by a roadside. He duly reported his find and investigations began.

The scenario for this tragedy opened at Marrakesh when Skytrain 4330719, together with 11 of its type, left there to fly to Cornwall via French Morocco on a delivery flight to the 37th Troop Carrier Group, 9th Troop Carrier Command. When they got to Morocco, three of the dozen

were grounded and returned to Marrakesh. The remainder pressed on and all but one arrived at their destination. There was no sign of the ninth so an over-water search was instituted. The aircraft had been carrying a cargo of new bicycles – the silent steed of the lucky infantryman. There was a US Army song at the time which ran: " . . . when a son of a gun in the cavalry, is travelling on a bike, what do we do in the infantry, we hike, we hike, we hike!" Or perhaps they were intended for bomber aircrew to cycle out to their planes which were widely dispersed on airfields.

Though people in the area had heard a loud bang several weeks earlier, this was ascribed to thunder and duly forgotten. This was early on the morning of Friday, 17 December, a day described as "the worst one that ever came, with frost, snow, and terrible showers of sleet and hail". The Skytrain had crashed into a ridge above Cummeennapeasta Lake – 'the lake of the serpents'. Once alerted, an army contingent started out at first light and many hours later came upon the scene of the crash enveloped in swirling fog.

Foynes flying-boat base welcomed three RAF Catalinas lost while on Atlantic patrols. This picture shows a day in 1943 when bad weather kept virtually all the civilian flying boats on the water – but not the Catalina which somehow managed to get into the picture!

An inquest was held on the five casualties before their interment which followed Catholic, Protestant and Jewish rites as the religious denominations of the crew could not be established. The possible causes of the crash included metal fatigue, structural problems, navigational difficulties, weather conditions, fuel problems and an unfamiliar route – this being the crew's first flight over the last leg of the journey. There was speculation too that when fuel was running low the crew may have been attempting to make for Rineanna. The winter weather in Kerry can be bitter and at an altitude of 3,000 feet the freezing air could have caused the propellers to ice up.

No one knows for sure what had driven the Skytrain into the mountain. There is little wreckage left now as local scrap dealers brought most of it down by mule – but a large portion of one wing subsequently blown down can be seen in the depths of Cummeennapeasta, its white star insignia still intact. At ground level, 'Casey's Yard' has a well-organised starting point for mountain climbers, where the young fliers are suitably remembered.

Another plaque carries the following warning:

Are you prepared for tough ground, wind, rain, cold, mist, accidents etc. You should have a map, compass, whistle, torch, first-aid kit, stout boots, protective clothing, food and drink. It is 100 colder at a 3,000 ft. summit – a breeze here is a strong wind up there. Take care on the way down – most accidents occur then.

It was a dangerous peak for an aircraft – and for rescuers too.

The US Navy PB4Y hit the 70-foot-high tip of Skellig Rock, burst into flames and crashed into the sea.

Ocean Liberators

It was not only as a bomber with the various US air forces that the B-24 Liberator made its mark. In the Battle of the Atlantic it was one of the aircraft which significantly changed the Allied fortunes when operated by RAF Coastal Command. The US Navy's version, the PB4Y, also became 'the scourge of the U-boats'. It was late in getting into the battle until President Roosevelt ordered a number of US Navy PB4Ys and Army B-24s to help close the Atlantic gap in March 1943. These included four USAAF squadrons of the 479th Anti-Submarine Group which aided Coastal Command's No 19 Group in their campaign to bottle up the U-boats in the Bay of Biscay.

The USAAF exchanged its ASV-equipped Liberators for an equal number of unmodified US Navy B-24D models in what became known as the 'great air horse-trade'. This swap allowed the Navy to build up Fleet Air Wing 7 (a unit based in Devon at a dismal aerodrome called Dunkeswell). Three PB4Ys came to grief off Ireland: one as a result of enemy action; one caused by a highly-improbable accident; and one beaten by the wild west wind. The first, on 26 February 1944, was pounced on by four Junkers Ju 88s about 140 miles due south of Skellig Michael, a 70-foot rock jutting up from the sea, eight miles off the Kerry coast. No wreckage or remains of the US sailors were ever located.

Later that same night another PB4Y was hunting U-boats which surfaced at night to charge their batteries, give the crew some fresh air and, perhaps, take on fresh fish from Irish boats which were prepared to sell to all comers. The aircraft struck the pinnacle of Skellig Michael, the impact

ripping open the fuel tanks. Its load of eight depth charges exploded seconds before the machine plunged down into 200 feet of water. Obviously, there were no survivors. But why was the aircraft flying so low? It would appear that the rock produced a radar 'echo' similar to that from a U-boat, and the airmen believed they were closing in to attack – but it was a phantom echo.

Earlier it had been noted that radar sightings of icebergs showed a 5-degree offset which persisted until visual contact was made. Perhaps in the excitement of impending action this lesson was forgotten? The day before the Skellig crash, another aircraft (63939) flying about 130 miles south of the Rock, crashed into the sea following an attack on an enemy – real or unreal? Another aspect of the collision with the Rock to be considered was that some, but not all, of Air Fleet 7's aircraft were fitted with a powerful searchlight and one wonders if this device had been in use?

While being ferried from Rhode Island via Iceland to the UK, another PB4Y (38799) ran into a belt of storm which had swept across the United States. The thin Davis wings of the PB4Y iced up; this was a common complaint – indeed it was often said of the Liberator's wing that "the amount of ice you'd put in a drink would be sufficient to kill its lift". Fighting the storm had increased fuel consumption and left the PB4Y with insufficient fuel to reach the UK, but with only two options a chance had to be taken; it seemed best to press on rather than to divert to Iceland. Within sight of the west coast of Ireland, the engines coughed and died and ditching was inevitable. The crew took to their rubber boat but this was overturned three times, with the loss of four members of the crew, most of the oars, and all the emergency rations and other equipment. The survivors fought on against a fierce north-easterly gale which kept blowing them away from land.

Eventually the boat reached the shore but of the four remaining, the co-pilot was dead from exposure. The pilot and a gunner staggered in darkness to the house of Michael Conneely, meeting a coastwatcher on the way. The latter used the pilot's whistle as a signal and he found the waterlogged boat and its four occupants, one dead and the others in a semi-conscious condition. He pulled the boat up the beach and helped the three survivors out. This burly Saygull took the weakest on his back and, telling the others to hold onto his arms, struggled to Conneely's house where there was a good fire, hot tea and brandy.

On the beach, another Saygull stood guard over the dead flyer's body overnight. In the morning a Protestant clergyman on holiday in the area borrowed a clerical collar from the local parish priest and performed the funeral rites. The man who died was the son of America's leading football coach and the US authorities were particularly concerned about his fate. After the service, his body was handed over with the usual impressive military honours to the US authorities at the border but the latter, as was usual, failed to provide reciprocal honours for their dead. In the meantime the survivors, under the care of the Sisters of Mercy, were recovering in Clifden district hospital. Initially they were so weak that they couldn't take food but eventually they improved, though the doctor in charge was still reluctant to release them. A trio of USN officers arrived to escort them to the border. Later, in the area where this aircraft had ditched, a PB4Y (serial 332211/E) sank the German submarine U-271. Perhaps this evened the score?

Lieutenant Joe Kennedy, the elder brother of the future president, served with Air Fleet 7 before being seconded with a colleague to Special Air Unit One. In June 1944 both men were flying a PB4Y (coded VB-105) which was packed with over 2,000 pounds of explosives. This experimental mission was coded 'Anvil' and its purpose was to blast German rocket-sites in France. When the two men set this 'flying bomb' on its course they would bale out over England while a 'mother' aircraft guided the huge missile by radio to its target. Unfortunately as Kennedy and his colleague were about to bale out, the aircraft suddenly exploded, killing both men. John F Kennedy, who was also serving in the Navy, therefore took Joe Kennedy's place as a potential presidential candidate in their father's plans.

In September 1997 a monument honouring the PB4Y crews who had perished was unveiled at St Finan Bay by the Irish Naval Association. Thirty-three next of kin attended.

12 'T'aint a Bird', 'Stinky' and others

"Is it a bird? Is it a plane?" queried the crowd gazing skywards in the Superman strip cartoon series. "Well, t'aint a bird" one of the characters observed. A Flying Fortress (serial 4230342) bearing the name 'T'aint a Bird' was to make an impact on the town of Clonakilty, an event commemorated in 1983 at O'Donovan's Hotel where the B-17 crew had stayed in April 1943.

In later years, Cristina Pisco, an American author living near Clonakilty, produced a successful romantic novel entitled *Only a Paper Moon* based on the incident. The arrival of the aircraft is dramatically recreated:

"Oh my God!" gasped Mary. It was heading straight for them. It was a plane all right. But it must be the biggest plane in the world. The screeching of birds could faintly be heard through the rumble of the engine. Mary flinched and covered her head with her arms as thousands of birds lifted off the estuary and flew overhead in front of the oncoming plane.

She was jolted off the cart as the pony reared. Tom jumped down and grabbed the bridle. The pony reared again, its eyes wide with fear; the roar of the engines matched only by the force of the wind. Mary held on to the side of the cart for dear life. She saw Imelda and Tommy struggle to keep the pony under control. Mary

stood as if suspended in time as the great green giant passed only a few feet over them and landed in the Marsh. Minutes passed and still none of them spoke. Mary's ears were ringing though the engines had been silenced. Out of the corner of her eye she saw Barry Kingston drop his scythe and start running towards the plane which was now directly in front of the causeway. Tommy let go of the pony and clambered down over the wall. "Where are you going?" yelled Imelda, hanging on to the nervous pony. "Tom, don't be a fool! Come back!" "I'll be grand," said Tommy, pulling at the brambles that were holding him back. "Look, the door's opening!" "You'll get yerself kil't," cried Imelda. "I will not get myself kil't! Don't be stupid." Tommy slid down into the marsh and ran towards the plane. "Why wouldn't ya?" said Imelda. "Come back! They'll shoot ya!" "No, they won't," yelled Tom over his shoulder. "They're Yanks!"

It was a good description of warlike arrivals in rural parts, though what actually happened was somewhat different.

As the B-17 came in, the pilots saw that they were about to land on swampy ground. They immediately cut all engines, hauled back on the controls to make a 'dead stick' landing, thus avoiding the distinct possibility of pitching over as the undercarriage touched. Three minutes later the

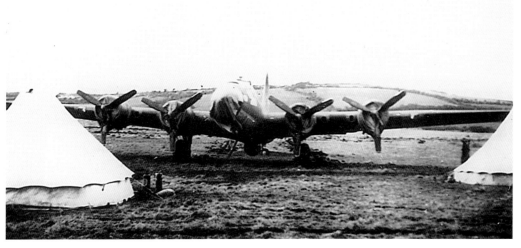

'T'aint a Bird' surrounded by the tents of the Field Engineers who are constructing a run-way for its departure. The aircraft's nose and engines are protected from the weather.

The crew of the B-17 'T'aint a Bird', together with an Irish officer, seen in front of O'Donovan's Hotel in Clonakilty. The airman on the extreme left is Guy Tice; second from the left in the back row is 2nd Lt W Prochaska who was to demonstrate the new bombsight; and on his left is Lt WK Thomas, the aircraft's captain.

Denis McCarthy

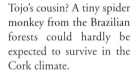

Tojo's cousin? A tiny spider monkey from the Brazilian forests could hardly be expected to survive in the Cork climate.

engines would have stopped of their own accord because there was no fuel left. When the aircraft rolled to a halt the co-pilot stuck his head out and asked Eddy Collins, who had been cutting rushes, if he spoke English, thus giving an indication that the crew was clueless about their whereabouts. On hearing a good-natured affirmative, the next question was "Where are we?" He was given an accurate answer – White's Marsh – which left the airman no wiser. The pilot now descended armed with a pistol and sub-machine-gun, telling Eddy, who was an LDF officer, to keep his distance until a finite fix was established. In hostile territory the B-17's self-destruct mechanism would have been operated, giving everyone just 70 seconds to get as far away as possible.

A more enthusiastic welcome greeted the crew when they reached the town of Clonakilty. The greatest interest there was in the crew's mascot, a spider monkey which they had been picked up in Brazil. Perversely they had named him 'Tojo', after the despised Japanese military leader. The

exotic little creature was as excited as were all the children gathering around to pet him. He had enjoyed the heat of Marrakesh, which had been the B-17's last fuel stop, but during a violent storm on the flight to Ireland, he had to be cuddled under the warm coat of one of the fliers. The tropics were his scene, not the chill air of west Cork, and he sickened. The local doctor diagnosed pneumonia, which proved fatal – it must be said the diagnosis relieved local residents who were afraid lest Clonakilty's famed black puddings, which the monkey had relished, had caused its demise. The LDF interred Tojo in the hotel's back garden and granted him full, though highly unofficial, military honours. He still lies there in the foundations of a ballroom which was built over the garden many years later.

The Fortress had come from Puerto Rico via Dakar, on a southerly flight which should have terminated in the UK rather than White's Marsh. 'T'aint a Bird' was intact but the 3rd Field Company of Army Engineers had to endure two weeks of constant rain while laying a half-mile runway

With engines revving, 'T'aint a Bird' prepares to leave White's Marsh on a specially constructed runway. *Denis McCarthy*

Sergeant Guy Tice is pictured with the commemorative plaque on the wall of O'Donovan's Hotel.

Irish Examiner

of Marston metal planking and bridging streams, if it was to fly out. When the runway was ready the great day arrived and everyone came out to wave a relief crew on its way to the UK via Rineanna. An Army report on the incident quite definitely states that this B-17 had initially been routed to North Africa but the flight had been cancelled in order to deliver as a top priority a new bombsight, and experts to demonstrate it, to the USAAF in the UK.

Subsequently the aircraft flew just one bombing mission during which it was badly damaged; when repaired, 'T'Aint a Bird' was used as a training aircraft. All the crew survived the war, though all but two ended up as POWs. One of the latter, Guy Tice, who came back to unveil a plaque on O'Donovan's Hotel, rekindled the time when he served in the 8th Air Force: "The average was 87% losses of aircraft; only 12% of crews were reported as coming back. One Group came back with one plane out of twenty; and we once came back with six planes out of nineteen."

The materials used for the runway were put to good use after the departure of the Fortress. The returning Tice was told by a farmer that his property boasted a large gate made from the metal planking.

'Stinky' and the top brass

'Stinky' was a B-17E (coded 419045) of the 97th Bombardment Group (BG) USAAF, prior to its squadron being re-equipped with the newer B-17F. The earlier types were reassigned to the 92nd BG, which was resting after a tour of duty. In mid-October 1943 two pilots of the 92nd were directed to take 'Stinky' to Burtonwood for modification, which entailed replacing the bomb bay with a huge fuel tank, fitting reclining chairs and a table, and generally refurbishing the aircraft up to the standard of a B-17G transport.

The two pilots returned after ten days to pick up their much altered aircraft; they were then ordered to fly it to Gibraltar with a full load of VIPs. 'Stinky' flew onwards from Gibraltar to Algiers on the second day of 'Operation Torch', the Allied landings in North Africa. When its passengers dispersed, it was assigned to General Carl 'Tooey' Spaatz, commander of the 9th US Air Force. 'Torch' progressed favourably as the top Allied political and military leaders met at Casablanca for a wide-ranging conference. Subsequently 'Stinky's crew were ordered to take Lt Gen Jacob 'Jake' Devers on the first leg of his journey to Washington via the UK to report on the eastern Mediterranean battlefields, and in particular the performance of US armour, in which he was well versed. His staff included one major general, two brigadier generals, a colonel, assorted lesser ranks, and a major as his air aide who, at Devers' behest, put the whole complement into jeopardy.

The pilots had chosen to make the flight via Gibraltar

While on its ferry flight to the UK, B-17 'Stinky' had to stop in Iceland for an engine change before going on to join the 8th Air Force. It subsequently flew on many operations before being converted to a VIP transport – eventually ending its days in a field in Éire.

US Air Force

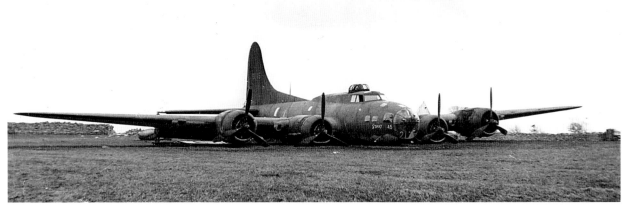

Lying low in 'the fields of Athenry' (Co Galway) in 1943 is the VIP Fortress 'Stinky' which contained General Devers and his staff.

Cdt J Ryan

across the Bay of Biscay at night in order to avoid the Ju 88 fighters which had been knocking down Allied aircraft in daylight. Forecasted weather conditions dictated that for most of the time the flight would have to be on instruments. The second pilot recalls:

> Just as we were turning onto a new planned heading, the general indicated that he didn't want to take the chance of crossing the Brest peninsula and ordered us to go west into the Atlantic and, in due course, turn back east for the UK. Our navigator objected, but very reluctantly concurred. Several attempts were made to obtain fixes from the RAF but it was busy controlling many aircraft returning from a raid. The radioman said: "In the midst of an electrical storm I got my butt chewed out right royally by the general who said, "What the Goddam Hell kind of radio operator are you? Why can't you contact England?" I offered him the headphones: he listened and suggested that I keep trying. The end result was that we arrived over Ireland, but we had no maps for

Another view of 'Stinky'.

this area. Turning for Northern Ireland, the fuel warning lights came on indicating about 20 minutes of flying time left, so we returned to the Galway area and I went back to tell the boss man that we intended to make a forced landing. Otherwise, I told him, we could take the plane up to 5,000 feet and everybody could bale out. He replied: "Son, what are you going to do?" I said: "Help the pilot to crash-land the airplane, Sir." "OK," said the general, "we'll do whatever you are going to do." On our approach to a grassy field our wheels were ripped away by a stone wall and we slid along to a stop. Just then I spotted smoke from No 3 engine. I somehow got out through a window (it's been a mystery to me for years how a big fella like me got out) and then opened the rear door for the rest.

When 'Stinky' had slid to a stop, children from a nearby school gathered around and I vividly remember tossing about a bushel of oranges to them – they hadn't seen any oranges for several years and I have never forgotten their excitement. Next a lovely old lady who owned the field showed up with the local parish priest and a gallon jug of whiskey. It being now midday and having had no food since the night before, the whiskey wiped away all our apprehensions. We were then driven to a local hotel where we had more of the same. By the time breakfast was served, I couldn't have hit the floor with my hat – a wonderful glorious feeling! The LDF had sealed off the building and after lunch we left through the kitchen door and set out for Northern Ireland by the back roads.

Shortly after the landing, a detachment from the 1st (Irish-speaking) Infantry Batallion surrounded the aircraft

At the centre of this group is Lieutenant General Jacob L Devers with, on his left, Major James Timoney, CO of the 1st (Irish speaking) Infantry Battalion.

Brendan O'Heagan

and took charge of its occupants and the considerable armament which it carried. As the Gaelic-speaking soldiers bustled about their business, the general was heard to remark, "Gee, these guys sure know their codes" – obviously thinking of American football. When he learned that his group would not be interned, he became quite chatty. The conversation was wide-ranging: in response to Devers' comments on US troops, the Irish infantry major responded that the morale of his own troops was very high; they were faced with the task of defending Ireland's freedom, recently won after a long and severe struggle within his own memory. He told Devers that the army's main trouble was lack of arms and equipment and the general asked him for specific urgent requirements, saying that he would see what could be done about this.

One of the group, Major General Brooks, had an Irish grandmother (what Yank hasn't?) and said he thought Éire had taken a very wise course in remaining neutral and should remain so as long as possible – as did his own

country. He added that he was very glad to have landed in 'Southern Ireland' and said nothing but good could come from his experience there. One imagines that such observations from senior officers would not have been welcome had they reached the ears of the US government.

Before the Americans parted from their escort at the border, General Devers complimented his host, saying that he had been very impressed by the general appearance of the Irish troops; they were the best turned out he had seen for a long time and he considered their uniform very smart and well kept. The Irish major said later: "I didn't tell him that I had been carrying out an inspection and therefore my troops were dressed in their best. He asked me to convey his favourable impression to all ranks of my battalion."

After-effects of 'Stinky'

There were various repercussions following the 'Stinky' incident. A letter from a Galway girl to a relative in the UK, which was intercepted by the censor, stated:

When operating as a bomber, 'Stinky' was captured in a wartime documentary.

Paul Browne

I believe they destroyed all their confidential papers when they discovered they were in Ireland, but they needn't have bothered. The army brought them into Athenry and got a meal for them at the hotel before suitable transport was arranged for a flying leap for the border. All our lads got oranges, bananas, champagne, lemons, and all sorts from them. Willy got a grand revolver and some sort of coat that inflates in the water. The Yanks could have been killed as there is a wood a few yards further on from where the plane stopped. The whole [of] County Galway has been arriving to see the sights. Paddy and Brendan arrived unexpectedly on last Tuesday, that's how we have all the first-hand news. I believe Hempel, the German minister and some more of the 'snakes' were down first thing next day, crawling all over the Fortress and spying out the land.

These last remarks were spurious, and Willy was soon relieved of the gifted revolver – as the Gardai repossessed all such souvenirs.

The Americans wanted 'Stinky' back despite being told that the inner wings were too large for road transportation and would have to be cut, thus complicating any rebuilding programme. In the construction of a B-17, the outer wings were attached to the inner wings, and these, in turn, were attached to the fuselage. It was further pointed out to the Allied liaison team that to get at these joints the outer skin areas, held by 7,000 rivets, had first to be removed. When 'Rosie The Riveter' (as a popular song of the time described

the thousands of women engaged on US aircraft production) punched them in at the Boeing works in Seattle, little did she realise that in the case of 'Stinky' these would have to be extracted in a field in Athenry in very cramped conditions. Under the skin, the joints were made by taper pins which required special extractors to get them out. Strangely, none of these tools were provided by the owners so the salvage men had to hammer out the pins with sledges over a period of four days.

Complete dismantling occupied a full month before the components were dispatched to Langford Lodge, the repair depot in Northern Ireland. Here it was decided not to rebuild 'Stinky'; if the Baldonnel boys had learned this, they might have been somewhat upset. Still, the operation once again proved their 'can do' attitude and, of course, 'Uncle Sam' footed the bill for their endeavours. The German minister knew very well of this ongoing breach of neutrality and made such a fuss that the crew of the next Allied plane to come down had to be diplomatically interned for a short period.

The second pilot on 'Stinky' had a final thought on the incident:

When we were back at Bovington via Northern Ireland I fully expected an investigation and a formal hearing on our 'SNAFU' ('situation normal, all fouled up') but nothing happened. I'm certain that General Eaker and General Spaatz realised that our crew had only a few months' flying experience, that this flight was only the third flight over water for our navigator, and that an investigation would be a waste of time.

There was an interesting sideline to this affair. When the Free French leader, General Charles de Gaulle, was due to return to the UK at the end of the acrimonious Casablanca summit meeting, his RAF transport was unserviceable and he was offered a passage in a USAAF aircraft. He declined, bluntly declaring in his usual forthright manner that he had no confidence in American navigators and didn't want to end up in his German-occupied homeland. So he waited until his British plane was restored to health. Perhaps he had a premonition of the 'Stinky' affair?

Film buffs will be interested in the next assignment for the two 'Stinky' pilots. The Hollywood producer William Wyler, together with a group of cameramen, had arrived to make the documentary *Memphis Belle* which many years

When the runaway Lockheed Lightning was finally secured it would have been towed through Belfast to Langford Lodge for degreasing and fully assembled for action.

later was the basis for a feature movie. The pilots were assigned to fly the moviemakers, who experimented with all possible camera stations in a B-17. Incidentally, in some of the sequences 'The Devil Himself', which had earlier spent 17 days on a Sligo beach, was to be seen.

The near loss of some of some of the Army's 'top brass' caused the US Signal Corps to establish a navigational beacon in the County Fermanagh border town of Belleek (world famous for its delicate porcelain).

Lightning strikes

The United States had been re-arming somewhat leisurely before the Japanese struck at Pearl Harbor. The pace of mobilisation quickened in all armed services and nowhere more so than in the Army Air Corps. Many senior NCOs, mainly staff sergeants, were now given an opportunity for pilot training – and could now sing with greater gusto the Corps song: "Off we go into the wide blue yonder, keep the wings level and true". One of these non-coms was Arthur L Brodhead, who did his training in sunny California before being promoted to 'Second Loot' and posted to the 95th Fighter Squadron of the 82nd Fighter Group, destined for the European theatre.

After a stint in Northern Ireland, the Group flew its P-38 Lightning twin-boom fighters over to St Eval in Cornwall as a jumping-off point for Oran in Morocco. In groups of four, the Lightnings took off following a twin-engined B-25

acting as navigator. Out over the Bay of Biscay they were jumped by a dozen Ju 88 fighters on their regular beat giving cover to U-boats going to and from the Atlantic. In the mêlée which ensued, Brodhead damaged one of the Ju 88s in a head-on attack and then downed another – or so he claimed.

After a hectic dogfight, Brodhead found himself alone in an empty sky and wisely headed back to St Eval. Disorientated after his first taste of combat, he soon became completely lost. His long-range drop tanks had been jettisoned in the combat, and with fuel ebbing from his main tanks, a landing became imperative. Luckily he spotted a beach whose grey granule sand absorbed an excellent wheels-up landing. He had come down at Ballyvaughan on the south side of Galway Bay where the LDF assisted him, somewhat dazed, from his cockpit and gently arrested him.

The part-time soldiers took Brodhead to the local Garda station where the police claimed him as their prisoner, but the captors, quoting the relevant section of the Emergency Powers Orders, quite rightly, and literally, stuck to their guns. The army arrived to collect Brodhead and thereby resolved the contretemps. While examining the unscarred plane, the author was questioned by a spectator who, looking at the single seat wondered, "What if the poor man was taken ill?" What indeed!

Another Lightning 'Shoo-Shoo-Baby', an F-5A

Left: Lieutenant Brodhead gazes from the cockpit of a P-38 Lightning fighter during training in the US.

Below: The peat bogs which were harvested by the Turf Board's machines were sometimes mistaken for landing strips by straying aviators.
Turf Development Board

photographic reconnaissance version which carried three cameras in the nose, was adrift over the central plain. The aircraft was low on fuel when the pilot spotted what looked like a landing strip. This was actually a bog where the turf was mechanically harvested to feed a nearby power station, a process which left neat lines which could easily be mistaken for well-camouflaged runways. Having circled for half an hour, he made an approach, snapping a telephone wire, but on touching down at 80 mph on the soft surface, the Lightning flipped over onto its back.

The watching turf workers, all members of the LDF, quickly released the inverted pilot from his straps and, in the words of the official report, "he was provided with stimulants and food". Restored, he wanted to speak to the US Embassy, being convinced that he would be let back to the UK which, in any case, was true for all USAAF personnel. He was of Irish descent and had served 18 months, the last two in England as one of the photo recce pilots of the 8th Air Force. These were known as 'Focus Cats' and though they were seldom in the limelight, their often dangerous missions supplied vital intelligence for planning the 'Mighty Eighth's' missions. The pilot truthfully explained that he had been on a training flight and that the 29 camera emblems, indicating that number of operational missions, on the nose of his aircraft represented the score of other pilots.

An Air Corps salvage team retrieved the Lightning via the bog's narrow-gauge railway and got it out on to waiting low-loaders. The locals claimed that this was the first time an aircraft had been recovered by railway, and certainly at national level they were probably right. The ubiquitous F/L O'Moore arrived and said he expected that there would be no claim for damages, which the RAF Air Attaché Malcolm Begg reiterated to army headquarters. No damage had been done, but both RAF officers were taken aback when the Turf Development Board put in a hefty claim for its workers' time.

At the beginning of 1943, a Lightning (AF4212802) featured in a truly touch-and-go incident. The aircraft, strapped to a large pontoon as deck cargo, was being shipped to Northern Ireland when in mid-Atlantic its vessel was torpedoed. The pontoon floated away, still with the heavily-cocooned fighter on top, until the composite cargo was spotted 500 miles west of Donegal and towed into Lough Foyle by a RN destroyer. Here a dispute arose between the Navy and the RAF as to who was to remove the aircraft – it surely can't have been a squabble over prize money? While the two parties argued, a huge tide caused the pontoon and its load to break its moorings and sail across to the Éire side of the lough.

The Air Corps was alerted and sent a party from Baldonnel with a crane, low-loaders and instructions to restore the plane and pontoon to Derry. The officer in charge was not delighted:

> The night I arrived in Letterkenny Barracks I was given a description of the job we would have when retrieving this undamaged and valuable plane ashore across rocks, mudflats, and seaweed – and I knew that I was in for a stinker. The army, incidentally, had mounted a guard at the spot to make sure nothing was pinched.
>
> When we reached the site next morning we found a bemused sentry regarding a bare strip of beach. He said that an even bigger tide had arisen during the night and had set the pontoon off on its travels once again. The Royal Navy met it again sneaking out of the lough and brought it back once more to Derry. This time, the plane was finally disembarked and sent to Langford Lodge depot to be stripped of its protective cocoon of black 'Paraltone' which completely sealed it from sea spray. This 'degreasing' involved hot paraffin, scrapers and scrubbing brushes – and

The relief pilot who flew this force-landed Tomahawk trainer back to base remarked that: "As a fighter, the Tomahawk was something of a rather rusty battleaxe!"

Captain A Quigley

> 40 man-hours. I was told later that this particular Lightning saw action over France and I am sure it must have been lucky, even though it gave me corns on my backside from travelling 300 miles in a lorry without even getting a glimpse of it!

County Monaghan, on the border, attracted a pair of Lightnings on 5 May 1944. One of them seemed to be in trouble and while its partner headed northwards it made a bad landing on Gormanstown aerodrome, breaking its starboard boom. The pilot explained that the pair had been on a training flight from Maghaberry and he was short of fuel. The remains of his unusual twin-boomed aircraft needed two low-loaders to get them back to the RAF at the border, north of Dundalk.

The total production of the Lockheed P-38 Lightning was 10,000 aircraft, made in 18 different versions. In the European theatre this long-range and versatile aircraft justifiably earned the Luftwaffe epithet 'Fork-Tailed Devil'.

More pugnacious Yankees

As well as the formidable Lightnings, other American fighter and attack aircraft occasionally appeared. A Curtiss Tomahawk serving with the RAF and bearing the name 'Bogey I' (coded AH920) came down in County Wicklow in April 1942. Compared to the Lightning, the Tomahawk was something of a pussycat (see above picture and caption). The initial Marks were poorly armed with two .30 inch guns and the RAF had relegated most of them to army cooperation duties in the Middle East or for training. The Wicklow example was a Mk I trainer based near Lisburn, County Antrim. The pilot blamed a faulty compass and fuel shortage for his arrival. He was removed to the Curragh internment camp, while his undamaged machine was taken to Baldonnel.

The new internee, a Flight Lieutenant sporting a moustache of magnificent proportions, was not enamoured of his new quarters. "Christ, I hope that I won't be in this bloody hole for long," was reported as his opening remark. He wasn't, because during the night the British Air Attaché spirited him back to Lisburn where his wife also lived. The unloaded guns on his trainer, indicating a non-operational flight, and a softening attitude towards interning RAF personnel, brought about to some extent by America's entry into the war, were factors in his immediate release. A relief pilot flew the aircraft back to Northern Ireland saying that as a fighter the Tomahawk was "something of a rather rusty battleaxe".

In February 1945 a B-26G Marauder (4468079) was en route from St Mongers in Wales to Liverpool, but somehow found itself instead over Gorey, County Wexford. With fuel low, the two crew members looked around for a suitable field but only waterlogged ploughed land was on offer. At that time of year the field was about to be sown with winter wheat as the maximum use was made of all arable land to provide 'tomorrow's bread'. On any surface a wheels-up

landing was bound to be heavy because of the Marauder's high wing loading – its limited wing area holding up its considerable weight. The B-26 was nicknamed 'the Flying Prostitute' because it had no visible means of support. Its pilots trimmed off most of the feathers on their silver 'wings' to underline this.

The salvage team leader recalled the mess his men faced:

The Yanks were short of spares for this type so we salvaged every fragment. To travel across the sodden fields we had to get army Bren-gun carriers whose caterpillar tracks and those of our cranes left trenches that looked like canals. There was mud everywhere and the scene reminded me of pictures of Flanders in the Great War. However, the job went well and we had some great parties at night. Rory [F/L 'Rory' O'Moore, RAF Northern Ireland] as usual was sailing around lubricating the wheels of endeavour with 'fusil oil' . He had to go back to Dublin halfway through the job and it struck him that our wives might be having a thin time while we were away mudlarking. He called my wife and those of other officers and gave them 'the time of Riley' at the Shelbourne and Gresham (then Dublin's leading hotels) for a few days. Years later, the girls were still bragging about the sumptuous time they had. We didn't worry if it was all on King George's expense account – he certainly owed it to us. I tore more uniforms, crushed more fingers, and burnt more flesh on salvage work than many a man who had been awarded the DSO.

Termonfeckin Strand, County Louth, 4 June 1943. The Martin Marauder whose 'nose art' surprised the nuns!

The officer in charge of dismantling Marauder 'Ridge Runner' with some pretty local girls. Unfortunately, or fortunately, the nose artwork is not fully visible!

The heavy impact of a wheels-up Marauder landing was again demonstrated when one put down on a firm flat field in County Cork. Its underbelly shattered and a large quantity of earth filled up the rear fuselage, no doubt helping to slow down the high-speed landing. Though the aircraft looked fairly intact at first glance, its back had been broken and it was a complete write-off.

Another Marauder startled a community of nuns who were relaxing in their holiday home in County Louth on a pleasant June morning in 1943. They were alarmed to see a war-weary Marauder swoop down onto the beach in front of their residence. Its nose bore the name 'Range Runner' and the tally of its bombing operations, as well as a dramatic depiction of a grizzly bear chasing an unclad young lady. As the crew sat sunning themselves on a wing, the hospitable nuns dispensed mugs of cocoa (which the men mistook for Irish tea), all the while discreetly averting their eyes from the artwork. No doubt, the holy sisters offered up some prayers for these hardy lads who went on to fly and fight another day – for they were quickly whisked across the border.

Soon an Air Corps team retrieved the aircraft's guns, warlike stores, and other valuable equipment. Less warlike were the condoms, lipsticks and nylon stockings with which the fuselage was stuffed. All items had been carefully labelled and loaded on a lorry, when a jeep appeared, the sole occupant of which paid scant attention to the salvaged materials. He saluted the Irish officer and said: "Hi, I'm

Above: This is the scene where the Norwegian pilot drowned when his Boston overturned in a bog.
Right: The intact aircraft at Montreal, showing its long-range belly tank.

Vincent Callaghan

Top Sergeant Charles Barnes, New York city. All I want, sir, is the Lootenant's dress pants, he's goin' to a dance tonite!"

A war plane circling for 20 minutes over Campile village in County Wexford was more ominous. The inhabitants were naturally uneasy because two Heinkels had behaved in a similar fashion before one of them bombed the local dairy factory in 1940. Three girl workers had been killed, and it was generally considered that the attack was not an unfortunate accident – the dairy's production was mainly for export to the UK. However, this intruder proved to be from the USAAF and its pilot selected a large field, known as 'The Long Run' (now the John F Kennedy Arboretum), for a belly landing. The aircraft skidded over the grassy surface but was badly damaged in the process. The official report of this incident ran:

> The crashed aircraft was a Bell P-39L Airacobra (Serial No 424518), a single-engined, single-seat fighter, painted dark olive drab on the upper surfaces and neutral grey on the undersides. The underneath of the fuselage was torn from the centre section to the rudder. The wings and undercarriage, retracted into the wings, were badly damaged as were the flaps. Two blades of the three-bladed propeller were crumpled up. The fuel tanks were nearly empty.

The salvage team leader added:

> The aircraft had stopped with its projecting 40 mm cannon pointing right at a cottage – it had a round in the breech which could have wrecked the joint. I sweated

bricks persuading this strange musket to disgorge its shell unfired. We then had to wheel it for a mile to the nearest road.

During his interrogation, the uninjured pilot admitted to G.2 Branch (Army Intelligence) that he had been on a delivery flight from Cornwall to North Africa, in company with others and led by a B-25, when the experience of the Lightning recounted earlier was virtually repeated – alone in bad weather, having to return to base, and coming down in Éire.

A ferry flight terminated prematurely in County Mayo on 25 October 1942. An A-20 Boston, coded BZ200, attempted a conventional wheels-down landing on what looked from the air like a good firm field but was, in fact, a 'shaking bog'. On touch-down, the aircraft flipped over onto its back, trapping the Norwegian pilot, Quartermaster Nils Bjorn Rosmussen, and drowning him. His two crewmen escaped unhurt except for some slight scratches. Local people said that immediately on touch-down the pilot realised his mistake and opened his throttles to get away, but alas without success.

A party of soldiers had to cut open the fuselage in order to retrieve the unfortunate man. The official report ran:

> Aircraft landed with undercarriage down which sank in soft, moving swamp or bog. Location of plane approximately four and a half miles from Crossmolina on the Belmullet road, then 500 yards down a lane to the north of the road [this is the nearest point to which a truck can be brought] and thence across another 600 yards of swamp to a moving bog.

An Air Corps engineer gave the following opinion:

> Very little further equipment can be removed, but it is possible that if the plane were to be cut in pieces a little more could be removed. A large amount of equipment was lost in the swamp and there is no hope of finding same.
>
> Radio gear not accessible for removal at present, probably OK. Removal of machine is practically impossible: there is no way of getting it out of the bog; the swamp is very dangerous and it would be useless to try putting down planks or sleepers as they would sink, in fact people could be very easily drowned during this operation.

The wrecked plane was donated to the farmer on whose 'land' it had crashed, but the night after the 1st Battalion guard had departed, sounds of sawing and hammering could be heard as it was dismembered by people from far and near. Within days the engines had sunk out of sight and there was little else left. Remarkably, some 60 years later when the crash is mentioned, many locals recollect seeing some part of the Boston, as a gate, a door, or some gadget which had been hammered into shape for use around the house or farm. Such were the exigencies of wartime shortages of material that it brings to mind the biblical reference, "What the locusts left, the mildew destroyed." But a constant reminder remains: the Norwegian's grave and its headstone are kept in good order to this day. One resident remembers:

As children we were always made by our parents to stop and pray for the dead pilot, and when we were finished praying, we cleaned whatever weeds and grass had overgrown the side of the grave. We were not the only family that did this, proof that in a small community like Eskeragh, people have respect for the dead, which was the case with this man, a person unknown to anybody.

Not all US aircraft could be called 'pugnacious': certainly not the pair that touched down in May 1944 in County Meath. These were L-4 Grasshoppers, a type which acted as aerial observation posts (AOPs) for infantry and cavalry, but principally for artillery. Their pilots said they were low on fuel when they landed which was soon remedied and away went the little green Grasshoppers. With the end of the war in sight, the manufacturers of the L-4 were obviously thinking ahead to the peace because their advertising copy at that time ran:

When victory is won it will be the peace-time version of the L-4 which will quickly and safely take you to your favourite hunting, fishing and vacation spots, saving you time, gasoline and tires. The L-4 will also streamline your business trips, saving you time, gasoline, and tires.

One wonders if the 'touch-and-go' pair survived and were 'civilianised', like many war surplus ones?

13 Fallen Fortresses

'Bugs Bunny', 'Bar Fly', 'Badger Beauty', 'The Devil Himself', 'Meltin' Pot', 'Stinky' and 'T'Aint a Bird' were some of the names chosen by B-17 crews for the Flying Fortresses which 'fell' into Éire. Some crews were lucky, others perished. Most were at the end of their tether after a strenuous crossing of the North Atlantic to add to the USAAF pool in Britain or to join specific Bombardment Groups of the 'Mighty Eighth'. A couple even arrived minus their crews, as will be seen.

The USAAF first tackled the 'Water Jump' in June 1942 with a trial trip by ten B-17s destined for Prestwick and for delivery to the 97th Bomber Group. The trial was unsuccessful and the formation broke and scattered, landing at various airfields. The reasons advanced for this failure were unforecasted bad weather, poor navigation, radio failure, and general inexperience. But by the spring of 1943 aircraft were being ferried over by better trained replacement combat crews, rather than ferry pilots. United States staff were now being attached to RAF flying control centres because some of the Yanks were unused to the

Thirteen B-17s are seen here after 'roll out' from the manufacturers. Nearest the camera is Serial No 124612, later to be named 'The Devil Himself'. It flew on many bombing raids and spent a month on a beach in County Sligo, before being returned to service and eventually flying back to the land of its birth for scrapping!

English accent and what they termed the 'Limey flying lingo'!

Rineanna played host to a total of ten straying USAAF craft, some of them Fortresses. Others of the type were claimed by the beaches, mountains and bogs which were not always inhospitable. During one autumn month there were at least 16 Fortresses circling in Irish airspace, totally lost. Brigadier General Edmond H Hill, commanding the 8th Composite Group in Northern Ireland, inquired if some navigational pinpoints could be established. During April 1943 he accompanied the Irish Chief of Staff, Lt Gen Daniel McKenna, on a tour of all aerodromes and emergency fields. The two discussed the question of marking each Irish headland where lookout posts were situated with 30-foot white numerals which would be illuminated at night. The Irish commander agreed and subsequently these signposts were marked on USAAF maps for crews which were now arriving in ever increasing numbers. In many instances these men, usually trained in the blue skies of the southern United States, said that this facility was "just dandy" for picking up their bearings.

Occasionally less fortunate Fortresses flew blindly into the mountains. Towards the end 1943, and in zero visibility, a B-17G (coded 4231320) flew directly into the Truskmore Mountains in County Sligo. Miraculously seven of the ten-man crew survived. This Fortress was up to two hours late on its ETA (estimated time of arrival) over the Irish coast, causing its pilot to descend through the murk to see if land could be spotted. All that remains today of this disaster is part of the aircraft's wing, now serving as the roof of a chicken shed. This aircraft was one of 40 Fortresses which had left Newfoundland for terminals in Scotland, Wales and Northern Ireland – many destined for the first two countries arrived at various aerodromes in the latter – one of them overshooting St Angelo and fatally crashing. Another, as far as can be discerned, ended up in the Atlantic.

From mid-1943 onwards more than a dozen Boeing B-17 Flying Fortresses destined for the build-up of the USAAF 8th Air Force in the UK came down. Some were being ferried across by the aircrews who would operate over Germany, others by ferry pilots delivering replacement aircraft. At the beginning of 1944 a B-17G (4231507) was hoping to land at Nutts Corner, Belfast, but the runway was blocked by another Fortress from the same ferry flight which had crashed on landing. One after another its four engines 'gave up the ghost', though some gave power spasmodically to keep the Fortress in the air. Eventually, eight of the crew were ordered to bale out followed by the co-pilot who had hoped to help in a belly landing near Enniskillen. The young pilot, Second Lieutenant Charles G Smith Jr, decided after the aircraft had stalled that it was time to go too, so he set the automatic pilot and baled out. Left to its own devices, the Fortress, with intermittent power, flew on. A strong north-westerly wind now took the crewless aircraft southwards until, its fuel tanks dry, it glided into a landing in County Kilkenny. It skidded across the fields in an exemplary wheels-up landing until it was arrested by stone walls and hedges; though damaged, it was far from being a total wreck. When local farmers dashed up they were astonished to see no sign of life but there was a vast array of uniforms, documents, tinned food, drinks, cigarettes and ammunition, with more of this 'loot' scattered around the nearby fields. 'Beachcombers' were soon active before local forces could cordon off the site.

An Air Corps salvage team was called in, retrieved the four valuable engines and returned them to Northern Ireland. Concurrently US Embassy officials arrived to discuss claims for compensation by the owners of the land while the Gardai conducted a house by house search and retrieved the cosy sheepskin flying gear which some of the locals were now sporting.

Though there were stringent checks on the personal kit of all airmen, both the quick and the dead, occasionally human nature and a soft guard commander, allowed items to 'go missing'. In the following case, Catholic guilt caused the girlfriend of one such to write to the US Embassy in Dublin:

Dear Sir,

A soldier friend of mine, who was in charge of a party guarding a crashed aircraft of the American Air Force, permitted members of his guard to take away some articles of clothing found amongst the wreckage. He now regrets this very much and knows that he must make restitution … He himself took away a pair of gloves, which he has since returned. In order that the other articles could be returned, he says that he would have to report the matter to one of his officers and have the articles taken from the men who took them away. He would probably get into serious trouble for what he has

Not all the USAAF planes came into Éire's safe havens for operational reasons. This Fortress is visiting Baldonnel to collect a supply of butter, eggs and bacon which were rationed at its base. On the extreme right is the second in command of the Air Corps, Major Patrick Quinn, with one of his aeronautical engineers beside him.

Tony Flanagan

done, and even then the articles might not be recovered . . . I promised to write to you for him and ask you to forgive him the debt, which he feels he owes to your government.

The letter continued in the same vein and the American authorities, well aware of the splendid job the army was doing, gave the boyfriend 'general absolution'! Despite this incident, security of official and personal belongings was a matter of strict procedure. The crew of another Fortress had been lodged in a local hotel but when they returned to their plane there were bitter remarks: "Half my kit is missing . . . So is mine . . . At least $100 worth of clothes they've had from me . . . Who is responsible for this? . . . What is the guard supposed to be doing?" The guard commander

politely pointed out that the airmen had been throwing out their bags as they landed and that all the missing items were in a neat pile close by. They were invited to check their gear and rather shamefacedly admitted that nothing really was missing. The repentant airmen also found a pair of shoes, some shaving gear, and a missing pair of socks which they had left behind in the hotel. Some very chastened Yanks were then escorted to the border in an icy atmosphere.

In the Kilkenny runaway incident, two young brothers somehow managed to get hold of some heavy machine-gun ammunition and decided to try it out in their father's double-barrelled shotgun. When discharged, the gun gave the senior youngster a mighty kick and split a couple of inches of the muzzles. The lads, in fear and trembling, brought the gun to the local blacksmith who cut off the

At the end of the war the USAAF offered numerous aircraft and equipment at knockdown prices to senior officers of the Air Corps. When they were returning to Baldonnel in this B-17 the pilot ignored instructions from the control tower and soon found himself entangled in the perimeter bushes of the base.

damaged two inches and apparently the father of the adventurous pair never noticed the shortening process.

A more lethal crewless Fortress which traversed the country diagonally created greater disturbance on 30 September 1944. This fully-loaded bomber had taken off from its Lincolnshire base when problems arose. Its autopilot was switched on and the whole crew baled out, leaving the aircraft to its own devices. A couple of Polish fighters tried to intercept it before it reached the Irish coast just above Dublin where USAAF Mustangs from Northern Ireland took over.

The runaway cruised along at 120 mph, just below the 1200-feet cloud base, helped by a tail wind of about 10 mph. It crossed the Midlands and when it was over County Mayo more Mustangs tried to pick up its trail as did two Air Corps planes, but the Fortress had about a 30-minute start on the latter. As the crewless aircraft met rising ground it miraculously followed the contours, maintaining its altitude thanks to the following wind. A headwind would undoubtedly have caused it to crash and one can only speculate on the outcome. Lookout posts on the north-west coast observed the runaway travelling out over the Atlantic

where undoubtedly its fuel ran out and it found oblivion.

This airborne version of the *Marie Celeste* provided a useful (though not sought after) two-hour-long practice for the military and civil defence forces – if one takes the long view! None of the pursuing aircraft – American, British or Irish – actually observed their quarry but many watchers on the ground did, little realising that an unpredictable flying bomb was overhead (see map on page 6).

In the village of Ballydavid, in the beautiful 'Kingdom of Kerry', a wall plaque commemorates the lucky escape of the crew of USAAF B-17 'Badger Beauty' (423279) which ditched off the coast in May 1944. This Fortress had been on a regular meteorological flight from a satellite field near its Bovingdon base, having stood in for the designated aircraft when it became unserviceable. These Met missions were flown to observe upper air conditions roughly an hour and a half before the 'Mighty Eighth' set out on a bombing mission. 'Badger Beauty' had covered about 400 miles of a 700-mile journey when one of its four engines began to smoke and had to be switched off. Not long afterwards, when the aircraft was two hours off the Irish coast, a second engine acted up and, because its prop could not be 'feathered' (to stop it spinning and reduce drag), the aircraft was now in a parlous situation. Warned over the radio of treacherous Mount Brandon, 'Badger Beauty', having jettisoned its massive long-range tanks held in the bomb bays, performed a perfect ditching five miles out. Both dinghies deployed automatically, but one did not inflate fully and was towed in by one of the 30 distinctive fishing 'curraghs' in the midst of which the plane had landed. 'Badger Beauty' sank 50 fathoms below the calm surface.

Medical aid and sustenance were provided but the only 'casualty' was the RAF weatherman who had sustained a twisted ankle. His seven US comrades, when dried out, went off to the local cinema in Dingle as if nothing had happened. Forty-five years later, the weatherman, who had been a meteorologist in civil life as well as in the RAF, returned to the little Kerry village to unveil a plaque in the bar. His words, delivered perfectly in the Gaelic of the region, were: "My principal gratitude for being alive is to God, but I owe you all a big thank you." He was presented with a fine model of the curragh which had rescued him.

Second Lieutenant Fred Rowan managed to get his B-17 'Bar Fly' closer to the shore than 'Badger Beauty' when he ditched at the other end of the country in Donegal. The aircraft, a brand new B-17G 20-VE, was being ferried to join 303 Bomber Group. Conditions were appaling and the aircraft had been hit by lightning in a severe electrical storm causing one engine to be feathered. 'Bar Fly' had not the power to fly above the storm and was ploughing through it when all the instruments went awry and the aircraft suddenly developed a spin. Fortunately, Fred was able to pull out of the spin when down to about 15,000 feet, but then trouble struck again: a second engine stopped, luckily on the opposite side to the first, and the Fortress somehow staggered on. To lighten the load, all personal baggage was dumped out of the rear door.

Just as fuel became critically low after 13 hours in the air, land was sighted; as Rowan put it: "I guess we all knew how Columbus felt when he sighted America in 1492". Seeing piles of burning 'kelp' (seaweed for use as fertiliser) the crew figured the fires were being lit for their guidance. As they closed on the Donegal shore there were three options: land on a nearby road, but it proved to be tree-lined; get down onto a beach, which was problematic; go back out for a

Second Lieutenant Fred Rowan, pilot of B-17G 'Bar Fly'.

mile, and land in the water and take to the life rafts. The third option was the one chosen: go out about a mile and come in again directly towards the beach at about 120 mph, cutting the engines at the water's edge and hoping to slide up the sand on the aircraft's belly.

When 'Bar Fly' was down to 15 feet, the crew discovered that the shallow water hid huge grey-green rocks, any one of which would have ripped the plane apart. Rocks seemed to be everywhere and could easily jam the escape hatches, but there was no chance of climbing away as the strand was surrounded by hills. At the last minute, Rowan pulled back on the controls to avoid the last rock and was soon down in the shallow water. Only one of the two dinghies operated so five got in and five hung on. Everyone was OK – but 'Bar Fly' had had its last drink.

At Mullaghmore in County Sligo, close to Lord Mountbatten's castle, a Fortress called 'The Devil Himself' and bearing the Bugs Bunny squadron insignia, made a conventional landing on the strand. It was en route from

England to Northern Ireland to cooperate with a Lightning Group in practising fighter escort duties. Weather conditions had been badly forecasted: in the words of its pilot, the operation was 'SNAFU' – 'situation normal, all fouled up!' Initially the Fortress only required refuelling, but when it was being taxied around on the strand to position for take-off, rough handling of the engines while on soft spots caused two to quit. What had started out as a day's work lengthened into 17 days after a salvage team dragged the Fortress away from the incoming tide. The men began to remove, refurbish and refit the engines (one new one was sent from Northern Ireland). Airworthy again, the B-17 now needed a 600-yard run to be cleared. During this long-drawn-out process, the Irish engineer officer – in the interests of further education – minutely examined the internal equipment; this included items such as a reflector gunsight and a highly-secret Mk VII Norden bombsight from which he noted that essential components had been dumped. He became very friendly with the crew to the extent that he invited one of them to his wedding the following year.

This incident provides an example of 'army accounting': four officers, three NCO's and 11 privates were involved, making a total of 19 days for the first, 40 days for the second, and 165 days for the privates. A truck and a mobile workshop were both used for 17 days, the former covering 926 miles and the latter 572 miles. This time includes the engineer officer measuring the beach and the medical officer standing by at take-off – not an item was missed. The rations and accommodation at nearby Finner Camp were also duly noted (though of course the salvage team would have to be fed and accommodated at their own base in any event). The damage to Mr Owen Mullaney's 'swamp' by the coming and going of the trucks was fully compensated by an agreed sum of £2 – the aircraft itself had touched nothing. In true army style, everything down to 5 lbs of mineral jelly was accounted for – and, of course, the wealthy USA had to foot the bill.

The Americans took their aircraft off into a useful 20 mph wind and made it back to base just in time for Christmas. 'The Devil Himself' went on many a mission and appeared in the original documentary *Memphis Belle*; it and its crew survived the war and returned to the America where the 'Devil' eventually got his due – like thousands of other Fortresses, it was scrapped.

A B-17G 20-VE, also bearing the 'Bugs Bunny' badge, looked for sanctuary on yet another Donegal beach in May 1943. Its pilot attributed his difficulties to a combination of Atlantic storms, headwinds, and a heavy load of aircraft spares which drained his tanks. Additionally, at the end of a 2,500-mile flight, problems with the navigation equipment and poor visibility didn't help: no-one could see whether they were over land or sea. The pilot, Harry Ford, recalls:

> We dropped into the water first to slow us and ended up on the beach. First on the scene was a group of school children who came charging over the dunes like a bunch of wild Indians. They seemed to be talking in a foreign language and I briefly thought that we had

A head-on view of Harry Ford's B-17G 20-VE beached on a Donegal strand after a difficult ferry flight.

Joe McLoughlin

Below: The first USAAF arrival at Rineanna was this DC-3 Skytrooper with a heavy passenger load of airmen destined for the expanding 8th Air Force in the UK.

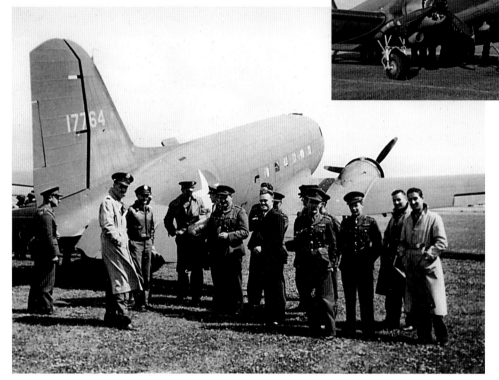

Above: Another view of the first USAAF arrival at Rineanna. In the very far distance can be seen one of the sleek Albatross civil airliners which formed part of the Éire–UK link.
Capt D Cousins

overshot our destination and perhaps they were Dutch, till I realised they were talking in Gaelic. Soon the Irish authorities had returned us to our UK base.

But the luck of this Fortress ran out six weeks later when it was shot down over Germany and only Harry and his co-pilot survived by parachuting into the same field, becoming POWs for two years. Fifty years later, Harry returned to Portnablagh with his son where he was given a special presentation of Irish crystal, a bolt of tweed and a framed photograph of Donegal Bay.

In recent years a Donegal sub-aqua club discovered a B-17 which had ditched in the middle of Lough Foyle. It had ended up there on a wartime ferry flight and the crew had been rescued by fishing boats. Despite 60 years' immersion, the Fortress seems intact. Its name, 'Meltin' Pot', can still be seen on its nose. Possible salvage and preservation are currently being considered.

Rineanna

Rineanna – its Gaelic name appropriately means 'Promontory of the Birds' – was selected in 1935 as the site for the Eastern Atlantic Air Terminal, which after the war became Shannon International Airport. Many USAAF stragglers availed of its concrete runways to the extent that on occasion it looked like an American base. The first transient visitor was a C-47 which was en route to Scotland. Its navigator had mistaken Northern Ireland as his destination, but further compounded this error by turning south and landing at Rineanna.

The C-47's arrival created something of a diplomatic precedent: though the aircraft was clearly not involved in warlike operations, it was carrying some of the first aircrews for the 8th Air Force. These 'combatants elect' were arrested and relieved of their side arms. Their release had to be authorised personally by Prime Minister de Valera, who also held the External Affairs portfolio: he concurred and the 'arrest' was lifted, whereupon conviviality broke out. Group photographs were taken before all retired to bed, but not before one visitor remarked: "I hope we get on as well with the English as we do with the Irish – but I doubt it!" The following morning, the C-47 was refuelled and provided with a new trailing aerial wire to benefit its communications. After a local test flight and an up-to-date weather briefing, it set out for St Angelo in Northern Ireland.

Nine other USAAF aircraft dropped into Rineanna. They were being ferried via North Africa to the UK to avoid severe Atlantic winter conditions. All had skirted the Iberian peninsula before all became completely lost. A Boeing B-17F (205220) came in on 21 March 1943, having circled the surrounding counties for several hours. A diplomatic precedent having been set, the Fortress and crew, enhanced by fuel and good fellowship, left later that day. A few days before Christmas, a B-17G (237895) was in a similar fix, or perhaps one should say lacking a fix. On landing, its captain declared that he was en route to Prestwick, but would say little else. His Fortress's 13 machine-guns were still sealed, confirming that it was on a delivery flight. An armed guard was posted, while the crew were fed and had an overnight rest before setting off with 452 gallons of fuel from Rineanna's limited stock. This would be returned, duplicated, later.

A pattern of islands in Clew Bay, County Mayo – an area constantly scoured by patrolling Air Corps aircraft.

Shortly afterwards, the first Liberator, a B-24H (252404) landed with the usual problems. When it came to a halt, the base fire engine was positioned at its nose to dispel any 'touch-and-go' ideas – but its exhausted crew simply asked to be put up for the night. When started up next morning, one of the Liberator's engines malfunctioned and the Air Corps set about trying to remedy the fault, but the recalcitrant engine caught fire and blew a fitter onto the apron, breaking his leg in the process. Four days later, by which time the crew were fully refreshed and very relaxed, a new carburettor was flown down from Northern Ireland in an RAF Airspeed Oxford, with three USAAF mechanics aboard. They worked all night before all the troubles were cured. Next morning, both Oxford and Fortress departed northwards, the latter having spent a week on the base. The next day, yet another wandering B-17G (231971) arrived: its navigator was given maps of Northern Ireland while the aircraft was being refuelled. Airborne he again missed his original destination and landed at an alternative airfield.

A fortnight passed before the first Marauder, a B-26B (295944) turned up. It had started out in Morocco and though armed, it carried no bomb load. It was refreshed with more of the precious 100 octane fuel and departed for Northern Ireland. The Marauder had barely taken off when a Liberator, a B-24J (2109825), hove in sight. Like the first, it too had taken off from Marrakesh. When it had landed, there was an explosion caused by the demolition of its radio equipment and bombsight by its security-minded crew. No other damage was caused and, full of fuel, it departed for the curiously named Nutts Corner, which became Belfast's civil aerodrome post-war – and a dangerous one at that.

During the winter of 1943–4, a pair of C-47s appeared, having flown around County Kerry trying to pick up their bearings. Their serial numbers were 224074 and 224098 from the same production batch. The usual ministrations were readily provided to their crews before the twins departed. On another occasion, with two USAAF aircraft concurrently on the ground, Air Corps personnel were surprised that neither crew fraternised with the other. The reason was that one was composed of 'Damn Yankees' while the other came from the Deep South. The American Civil War was still remembered on the green grass of Ireland, even during a worldwide conflict!

Towards the end of 1940 when Germany was at the height of its conquests, its foreign office decided to send two German army officers and a radio operator to Éire in the guise of consular officials, pointing out to its representative that Dublin was now regarded as an important observation point for surveillance of British military and maritime activities. The soldiers, who were named, would be flown to Rineanna, but to underline the innocent nature of their duties they would travel in a Deutches Luft Hansa airliner D-AGAK. The Irish authorities were asked to give details of available radio beacons and frequencies but de Valera put his foot down and ordered that any such aircraft should immediately be arrested by ground forces. Von Ribbentrop, the German Foreign Minister, was profoundly annoyed! The Rineanna garrison stood to and on Christmas Day observed a German aircraft – but it flew on without attempting to land.

Rineanna was the link for passengers and cargo on UK landplanes and the transatlantic flying boats at Foynes. Travellers were bussed the 40-odd miles between the two bases, one on the northern shore and the other on the southern bank of the great Shannon estuary. DC-3s, civilianised Whitleys and Hudsons and the beautiful four-engined De Havilland Albatross airliners operated connecting flights to and from the UK and Rineanna. The sleek DH Albatrosses were used initially and 'Frobisher', 'Fiona' and 'Fortuna' (respectively G-AFDJ, G-AFDM and G-AFDK) were constantly to be seen on the apron. Seven of these elegant craft, all of wooden construction, had been built, the first two going to the RAF and the remainder taken up by BOAC, as Imperial Airways had become in 1940. A version of the Albatross had originally been intended for transatlantic duties and, as events proved, it was just as well that this scheme was abandoned.

In August 1942, 'Fiona' was on a return flight to the UK terminal at Whitchurch near Bristol when a large section of its wing root fairing flew off, causing violent vibration. The aircraft returned to Rineanna where the Air Corps effected the necessary repairs. However, this incident was a portent of worse to come when, some months later, 'Fortuna', on its approach after a flight from Bristol via Hendon, began to break up in the air. The pilot descended rapidly and the aircraft disintegrated further when it hit the ground – luckily missing Rineanna's only pub! The passengers included four members of the BOAC board and four

Security was strict at Foynes – locals applying for jobs there were vetted by the Garda to intercept would-be saboteurs and, as seen here, an armed policeman guarded any flying boat not moored out in deep water.

executive officers as well as five crew members. None of them were seriously injured. 'Fortuna' had literally come unstuck and its two sister planes were immediately scrapped. The wreck was examined by Air Corps engineers, one of whom described its condition:

> The plywood skin of the starboard wing had been replaced with insufficient glue and it simply tore off when the flaps came down during the final approach. The port flap, of course, remained fully down causing the aircraft to yaw through 180 degrees. The most noticeable additional element was a violent odour surrounding the site which we discovered emanated from the only casualty – a goat buried under the fuselage. My word, were we glad when we had finished that job! One of the eyewitnesses, an old lady who lived in a nearby cottage, told us that she thought the aeroplane had been distributing leaflets but found that these were pieces of plywood. She was an exemplary witness and her contribution to the investigation was most helpful because we found not a trace of glue on any of the pieces she gave us.

The Albatrosses were replaced by civilianised AW Whitleys whose crews, like others using Rineanna, became somewhat uneasy as the barrels of AA guns guarding the base followed all aircraft on their approach – this was by way of practice of course, for Rineanna was truly a safe haven for all comers.

After the European conflict its human wreckage – badly wounded and traumatised American troops – were airlifted home via Shannon, as Rineanna was now known – a pitiful scene soon to be replaced by happier travellers.

14 Clipped wings at the Curragh

The British army established a training base on the plains of County Kildare in the middle of the nineteenth century. In time it grew into a complete military town – the Curragh Camp. The new Irish national army in 1922 took over the complex of seven red-brick barrack blocks, married quarters, a hospital, and good sporting facilities which included a full-sized swimming pool. Close by was the old Royal Flying Corps landing ground. During and after the Irish Civil War, many of the anti-Treaty forces were interned there, as were their successors, the 'new IRA' in the 1940s. They were held in 'K Lines', better known as 'Tintown' from its acres of corrugated iron cladding.

Here in new premises, downed airmen of both the Royal Air Force and the Luftwaffe were interned ostensibly for the duration of the war, but in the case of RAF personnel, a benevolent system of early release was practised. Both nationalities kept the traditional distinction between officers and other ranks, but the Germans of all ranks, at

'K Lines' – the internment camp at the Curragh as it looked in 1942. B Compound for the British is on the left-hand side, back-to-back with G Compound for the Germans.

Luftwaffe and Kriegsmarine officers interned at the Curragh are seen here in 1944.

their own request, ate in the same mess. The first of these was the crew of the Condor which had a miraculous escape on the heights of Mount Brandon in August 1940. The first British arrival was an unlucky fighter pilot who had run out of fuel while over County Waterford that September. Gradually both internment camps began to fill up until there were some 140 Luftwaffe and Kriegsmarine men side by side with 100 men of various nationalities in the British camp. Successful escapes and benevolent repatriation diminished RAF occupancy as the war progressed.

The barbed wire which encircled each compound was guarded by the resolute corporals of the Army Police Corps, distinguished by their red-banded caps, who answered to either the initials 'PAs' (short for Poilini an Airm) or simply 'Red Caps'. The overall 'landlord' of the internees was the Officer Commanding, Curragh Command, Colonel Thomas McNally, though the day-to-day responsibility was left to other officers.

An accurate account of conditions for internees was given by F/L Jack Calder, a peacetime journalist who wrote a number of stories for the Canadian press. An extract from his first endeavour, published in 1942, runs as follows:

Thirty-four perplexed young men with hundreds of thousands of dollars' worth of flying experience, are almost living the life of Riley in the British internment camp in Éire – almost, but not quite. They are Allied airmen who, through one reason or another, have made forced landings or bailed out over neutral Éire or have come down in the sea within three miles of the coast. The personnel comprises Englishmen, Scotsmen, Canadians, Poles, a Welshman, a New Zealander, a Fighting Frenchman, an American and even a wireless operator from Northern Ireland. The American is F.O. R.L. Wolfe, a member of the now defunct Third Eagle Squadron, and a good few of us have Irish blood.

Almost as much nonsense has been written about this camp as about the neutrality of Éire. For these young men, all of whom joined voluntarily in the fight against the Axis, it is a situation of paradoxes and puzzles, of friendliness and bitterness under conditions that none of us sought and all would do anything to end. We travel about the countryside on parole within set boundaries and freely discuss Éire's neutrality in public houses and private homes. Once a week we are allowed to leave our camp on the Curragh for a one-day visit to Dublin, 30 miles away.

Some of the world's best horses are bred and trained on the Curragh and one of the world's best-known race courses is only a mile from us. Outside our working time, we find diversion in going to the races, riding, swimming, fishing, football and walking or cycling through the green countryside. Three of the internees play rugby for neighbouring Newbridge, three of us have played soccer for the same town, and our own soccer team plays Irish Army elevens regularly.

So friendly have we become with the horsemen that racegoers frequently approach us for tips. So friendly are some Dublin merchants that we are able to maintain a first-class bar in the camp. We may bring in as much food as we can buy. It would be a luxurious way of living through the war, for we remain on full pay. Those who want to may study for post-war activities.

Yet the main topics in the messes are war and flying – so much so that a disillusioned airman will walk out of a discussion to read or go to bed or lean on the bar. At breakfast someone always relates a dream about escaping, or flying, or home. Unfortunately this is no place to discuss escaping or escape attempts. My sorriest failure came when my great friends F/L Grant Fleming, DFC, of Calgary, and F/O Bob Keefer of Montreal got away last August, subsequently returning to Canada.

Since then miles of barbed wire has been added to the several fences. The other main obstacles remain the same: raised blocks which prevent tunnelling, a small camp area and a force of intelligent guards handpicked from the Irish Army. They know a thing or two about escaping as a result of their experiences in the 'troubles'. There has been one recent consolation for our plight. Often we used to have to listen to "Deutschland Uber Alles" and shouts of "Heil Hitler" from the German camp on the night of Axis victories.

When the start of Montgomery's march in North Africa was announced, we smuggled fireworks into our quarters. The young Nazis have never had a chance to reply to the hullabaloo we raised that night. One thing, more than any other, makes us writhe. That is to have someone say: 'You ought to be glad you're safe. Your people must be very happy.' We realise that we are safe and that our relatives must get a certain satisfaction from that knowledge. But the allusion makes us want to break

The Luftwaffe and Kriegsmarine band in the Curragh Camp.

"I've been here so long that I'm almost a Catholic, and if the Irish continue to plaster this bloody camp with blasted barbed wire I'll be staggering out of here a teased out old bastard with a bloody Irish brogue!" Remark attributed to P/O Aubrey Covington after 16 months in the Curragh Internment Camp.

down the barriers of diplomacy on which we have been so carefully instructed.

Shortly after this article was published, Jack Calder ceased to go out on parole and spent most of his time in his room writing a book. He gradually became sullen and bitter and more and more of a recluse, but this behaviour was a well-acted ploy to get free. One day the Red Caps found him writhing on his bunk with his lips and mouth stained with iodine and an empty bottle clutched in his hand. As he was being brought to the hospital he made great play of vomiting and a fellow officer made the driver stop. At this Calder jumped out; he was wearing his running shoes and carrying a haversack and he sprinted off. Recaptured, he kept up the pretence of illness and was transferred to an Irish hospital and later to a London one where he had

This home-made Christmas card shows that 'goodwill to all men' was the order of the day in the internment compounds.

Another example of the Christmas spirit at the Curragh was displayed in this letter which runs as follows: "Dear George and Rudy: We would like to thank you for your beautiful cards and your thoughtfulness in wishing us Christmas Greetings. May we wish that you also may have a very merry Christmas. Your two comrades in the RCAF, Paul Webster and Charles Brady."

great difficulty in proving that he was in fact sane.

Arthur Voight, previously encountered when he survived his Heinkel force-landing in Wexford, also set down his impressions in a 'jail journal' from the point of view of Luftwaffe internees. Voight had suffered a harsh regime when he was a prisoner of war in Belgium and France during the Blitzkrieg of 1940 and he was not looking forward to further incarceration. But right from the beginning his fears were allayed, as the following extract from his memoirs shows:

A Canadian flight-sergeant pilot and his Irish bride after their wedding. Most of the guests seen here were fellow internees at the Curragh.

Sergeants Three: these smart RAF NCOs interned at the Curragh are all set for a social event.

On arrival at the Curragh we were led into an ante room, whose comfort was unheard of in any prisoner-of-war camp – internment must be different from what I could see so far. Handshakes and general introductions, followed by discarding our battle dress for the last time and we made ourselves comfortable for a very long chat with our eleven comrades. We found out that they were better informed about the war on all fronts and in fact they knew more than we did because of their big modern German Blaupunkt radio. Their main occupation was listening to all news and comments, as they had little else to do. Another thing, something that would be very risky in Germany, they often tuned in to a British radio station to hear the enemy's version of the war.

Our crew was the third German one to be interned. The first one, a 'Condor', came down in Kerry near the village of Brandon, with two men injured, not very seriously. The second one, a flying boat, had to ditch near a small islet with an unpronounceable name for a German tongue [Innisvickellaune!]. The senior officer, Oberleutnant Kurt Mollenhauer, was commander of the Condor whose pilot was Robert Behrems [who] had some pre-war experience over the South Atlantic with Luft Hansa airline.

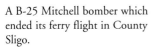

Luftwaffe NCOs in the Curragh internment camp.

A B-25 Mitchell bomber which ended its ferry flight in County Sligo.

Sligo County Library

Robert told me much which at first seemed almost incredible for us newcomers in a neutral country. I asked him what time was breakfast? "From nine o'clock on, you may be surprised, but don't forget the sun here rises one hour later than on the continent." "What are we getting for breakfast? . . . A cup of coffee perhaps and jam?" He burst out laughing, turned around and told the others what I had just said. "No, we do better than that, you are forgetting this is a neutral country untouched by the war. We have tea, not coffee, bacon and egg if you like, jam or marmalade and porridge." "Porridge, what's that?" "Oatmeal flakes", Robert replied. "Oh, I remember we had that as children, if it could be got at all after the First War." Robert

reminded me that we were not prisoners of war but internees like our neighbours. "Like what neighbours?"

"The Royal Air Force men on the far side of the high dividing wall, which you might have seen." "You mean there are British airmen in the other side of the camp?" "Oh yes, almost as many as us." "How did they get here?" "The same way as we did, crash-landing, ditching, swimming ashore, or baling out." "But dammit, we covered between three and four hundred miles over the Atlantic and had to force-land here, because our machine was a bomber and not a glider from an aero club." Robert replied: "You have a good point there; we too were surprised to find so many British airmen here.

Come to think of it, from Wexford to the nearest spot on the French coast is a few hundred miles of water. From the Irish coast it is not much more than seventy miles to England." "And they have been interned like us, despite the fact that they are British and virtually next door neighbours?" "Yes, strict neutrality is the motto here," he confirmed. Oberleutnant Mollenhauer came up with a suggestion worth thinking about: "What you need here most is a change of mentality." It was now early in the morning and we decided to go to bed and continue to explore our new surroundings tomorrow. Richard and I shared one room, Max's new roommate was the rear gunner of the flying boat, in the opposite building. Kessel got a room near the other officers. Showing us our room Robert said to us: "You can manage tonight without, but tomorrow you can get your pyjamas and toothbrushes . . . good night, schlafen sie wolh". Sleep well, he had said, and I certainly did.

Through various ruses and semi-successful escape plans, the RAF quota at the Curragh diminished. The big escape of 9 February 1942 caused a full-scale 'punch up' between some of the RAF and the Red Caps, the latter employing batons and blank ammunition and also firing some live rounds over the heads of the escapees. Choice language was exchanged between both sides, some of the veteran Red Caps having been imprisoned by the British 20 years before. The first internee to be recaptured was a Polish sergeant-pilot, Stanislau Karniewski, who tried to bluff his way by telling his captors in broken English that he was a butcher coming home from a night shift, as borne out by his hands and feet which were a mass of cuts from the barbed wire. The soldiers were highly amused because Stan, as he was always known, was popular with everyone in contrast to his fellow Pole, Captain Kazimierz Baranowski. The latter was more than 40 years old, and because of his very limited knowledge of English, kept to himself. His impeccable manners and the distinguished air in which he smoked a cigarette from a long holder, called everyone by their surnames, drank only cognac, wore a hairnet in bed, and presented such an aristocratic impression, earned him the nickname 'the Baron'. He treated Stan as his batman. Whenever 'the Baron' appeared, Stan would spring to attention, salute him and if the autocrat directed him to polish his elegant shoes, would respond, "Yes sir, right away sir."

The urge and indeed the duty to escape was a continual

This series of photographs shows some of the 'hand-over' points on the Northern Ireland border where Air Corps transport returned aircraft and equipment. The top photograph shows, on the right-hand side, the unmistakable remains of a Wellington's geodetic fuselage. The middle picture shows one of the 'Queen Mary' low-loader which has just disposed of its load. The third photo is of a typical customs post with an army station wagon behind the barrier.

Captain P O'Shea

preoccupation in the internment camps. It represented perhaps a sadder side of an otherwise easy regime at the Curragh. Being separated from one's family, friends and country cannot be counted as wholly satisfying. Many of the RAF escapees got over the border and back to their squadrons: sadly this dedication later resulted in death, either in action or in accidents.

Lone eagle: French cuisine

The RAF's Eagle Squadrons were American pilots who had volunteered to join in the fight before their country did – in much the same way as 'The Flying Tigers' assisted Nationalist China. One of these eagles was Ronald 'Bud' Wolfe, a 22-year-old who was returning to the 133 Eagle Squadron base at Eglington on the last day of November 1941, flying Spitfire B8074. He had been on convoy patrol when his engine quit because of lost coolant and minimum fuel. He kept the engine going for a few minutes intermittently. To add to his problems his radio had failed – he could only transmit and could not hear instructions from base.

Knowing he could not survive long in the sea, he headed for the nearest land and baled out, seeing his flying boots taking off in formation as the shock of the parachute opening shook them off. Floating safely into Éire, he was met by a local girl who promptly invited the tall, gallant and glamourous fighter pilot to dinner. He declined because he wanted to check on his crashed Spitfire and get over the border, but failed in both endeavours when he was taken into custody.

He did, however, dine well, courtesy of the occupants of B Camp when he was installed there. One of his hosts recollected:

> Osberstown House always lifted our morale and we hoped it would cheer up our guest. I am sure it did because it took us nearly an hour and an eight-mile bicycle ride for us to make a reasonably sober entry into the camp – to the smart salutes of our captors, Military Police corporals. A week went by, and late one night our radios picked up a newsflash from Hawaii where it was early morning on 7 December 1941. We remained glued to our radios, speechless and thunderstruck with the reports of what Roosevelt was to describe as this "day of infamy". Initially, we were stunned by the news that the Japanese had clobbered the American fleet at Pearl Harbour; but when we realised the consequences, we all rushed to the bar, slapped each other on the back, shook hands with our new ally and began to celebrate what was certainly the best news since the outbreak of war. We

turned our radios on full blast, sang and shouted, and this time it was our turn to keep the Germans next door awake all night.

But the new situation put Wolfe in a fix. He maintained that as a member of an Eagle Squadron, composed of US volunteers, he had not sworn allegiance to Britain and was merely a mercenary who, if released, would guarantee to take no further part in the war. He said that he would now be able to continue his civilian flying profession, but if he continued to be incarcerated his standards of professional piloting would deteriorate. Now that the United States was at war this scenario was no longer valid.

After one a week of internment Wolfe went missing and shortly thereafter reported back to his squadron contending that he was a legitimate escapee. The Yank had signed a parole slip and left camp but immediately returned on the pretext that he had forgotten his gloves and went off again. Wolfe maintained that by re-entering the camp he had automatically discharged his parole and had not been asked to sign another form before re-exiting. In the event, he was adjudged by the British authorities to be in breach of parole and was returned. The American minister, David Gray, arrived at the Curragh and, though no great lover of Ireland, agreed that the internees were enjoying complete freedom of movement under a very liberal parole system – but it was natural that young men would wish to get back in the war.

However, Gray acknowledged that the commander at the Curragh would feel that his goodwill had been ill-requited by a trick that infringed on the spirit of parole.

The matter remains a bone of contention to this day, with Wolfe still maintaining that he was completely in the right. Whatever the rights or wrongs, he was released, together with most of the RAF aircrews, on 18 October 1943 and transferred to a USAAF P-47 Thunderbolt squadron at Duxford. He stayed in the air force after the end of hostilities – not quite in civilian aviation as per his earlier special pleading.

When the owner of Jammet's, Dublin's famous French restaurant, heard that another non-British pilot, Maurice Remy, was interned, he wrote to the authorities as follows:

> Dear Sir,
> I have just learnt that an Air Force officer of the Free French Forces has been lately interned in the Curragh Camp. I would be deeply grateful if you would let me

Buglers sound the last post and infantrymen present arms at the obsequies of a Luftwaffe crew, as for all fallen airmen during the war.
John Scanlon

know if it would be possible to get in touch with him, if I may get him things he might require, or perhaps visit him. Would there be a possibility of having him out on parole for a few hours?

About the time the letter was received, regulations permitted one visit to Dublin per month and no doubt the French pilot enjoyed many samples of his country's cuisine. Dublin is the poorer for the disappearance of Jammet's.

Procedure and aftermath

When a warplane crashed or force-landed, the first priority was medical attention if needed. Otherwise, intelligence officers conducted their interrogation of downed airmen prior to refreshment and rest being granted. It was felt that a more accurate story could be obtained while the survivors were still recovering from their mishap. The Chief of Staff laid down that a specific amount of liquid hospitality be offered to survivors and added:

"Each man may be supplied with not more than twenty cigarettes per day in the interval from the time of force-landing or crash until placed in an internment camp." RAF crews, depending on the political situation and particularly in cases of those injured, were whisked across the border.

Four border towns were the usual handover points for aircraft, equipment and the injured. When the Allied authorities requested that the dead should be handed over at the border, this was done with due ceremony. The usual Irish guard of honour consisted of an officer, 16 men and a bugler. Most of the handover ceremonies at the designated four border towns involved the 17th Infantry Battalion whose area covered Donegal and part of County Sligo. A reciprocal guard of honour was formed by the RAF; it was unarmed because appropriate honours would be rendered at the obsequies. The USAAF was much less formal in the treatment of its deceased to an extent which local people, with their traditional reverence for the dead, deplored.

While the bulk of salvage work was undertaken by the

Air Corps, assistance by the RAF in the form of technicians and equipment was occasionally provided from maintenance units in Northern Ireland. This sometimes created problems when young men were billeted in a town where there was likely to be competition for the local lassies from Air Corps and local civilians at dances and other social occasions.

This competitive situation arose when an RAF B-25 Mitchell on an ferry flight made a forced landing, through faulty navigation, near Roscommon town. There was some difficulty with dismantling the centre section of this type, so the assistance of an RAF maintenance unit was enlisted. An elderly lady, then a young girl, reminisces:

> The first to arrive at the plane were my brother Stephen and Tommy and Michael Shaughnessy, with me a close fourth. We looked in awe as two men stepped from the cockpit and gave us chewing gum. We had a great time looking around the plane until the army arrived. People started to arrive from all over the county, most of them on bicycles and we enjoyed watching all the excitement. Twelve good-looking young men arrived from England and pitched their tents in a 'fairy fort' that was sheltered by trees. In those days there was very little by way of diversion and the older girls were wined and dined by the English boys which didn't please the local lads!
>
> My mother cooked roasts for them and gave them milk and in return they gave her paraffin

The German cemetery in County Wicklow where all the German air dead were reinterred in 1959.

which was like gold dust because of rationing, so our house and our neighbour's always had a light in the window because my mother always shared the paraffin. Romances flourished; we used to carry messages from the lads to the girls in the village and vice versa. No marriages blossomed out of it but it was a time in our lives that we will never forget.

Fifty years after the event a granite memorial, which featured a silhouette of the Mitchell, was erected close to where it had landed. The pilot and his son had come from the United States and joined British and US Defence Attachés, local politicians, and Irish officers. There was a dinner in the evening and a special parish welcome which included Irish dancing, stories and music. This was Percy French country and no doubt 'The Mountains of Mourne', 'Abdul-Abulbul' etc got a good airing! Incidentally, with true Marian devotion, the last line on the memorial reads: "Queen of Peace, pray for us."

Aftermath

There were 223 fatalities among the 800 airmen involved in crashes in Éire. This number included airmen whose bodies had been given up by the sea. All the dead were treated with the greatest reverence and the funerals of those buried in Ireland attracted many mourners. In the old tradition of those times, window blinds were drawn and villages and towns came to a standstill as the corteges proceeded, coffins draped with the appropriate national flag and flanked by military pall-bearers. Ecumenical services were conducted and full military honours rendered at the gravesides.

The Allied casualties, mostly RAF, were, in the main, interred in local Church of Ireland grounds. Some were repatriated to the United Kingdom, but mostly the airmen rest where they were initially buried. Unlike the Germans and the Americans, the British survivors rarely held reunions.

Shortly after the end of hostilities, the new US President, Harry S Truman, proclaimed 1 August as 'Army Air Forces Day'. The US minister was host at a dinner in Dublin and though he had been strongly antagonistic to Irish neutrality throughout the war, he now expressed the gratitude of the American people to the army and the Air Corps for their "unfailing kindness and tenderness to 'shipwrecked' airmen, whose wounded had been cared for with solicitude and whose dead had been buried with the greatest respect". The Chief of the Irish Defence Forces, Lt Gen Daniel McKenna, responded: "I am glad that Ireland attended to the downed airmen in the way they deserved, and rendered to all airmen, of all nationalities, the assistance and honour which was their due."

The German survivors formed an association of Irish internees in 1956 but with the passing of time their numbers have dwindled and the association has lapsed. In 1959 their dead were eventually reinterred in Glencree, County Wicklow, a place of rare beauty and serenity. There, a plinth is inscribed with a poem in Gaelic, German and English, whose sentiment could apply to all the brave airmen who found their last landfall in Ireland:

It was for me to die under an Irish sky,
There finding birth in good Irish earth;
What I dreamed and planned, bound me to my
    Fatherland,
But War sent me to sleep in Glencree,
Passion and pain were my loss, my gain;
Pray, as you pass, to make good my loss.

List of Incidents

Unless noted, all the incidents hereunder were crashes involving substantial damage or total destruction. 'Force-landed' indicates a controlled landing, at times a normal landing, with the aircraft intact or slightly damaged. This list makes no claims to completeness because there are still 'grey areas' surrounding some listed, and indeed unlisted, incidents. Where known, Serial Numbers are shown in brackets and Squadron Codes, as used at the time, between quotation marks. Dates occasionally vary within a few days due to differing sources. A small number of incidents outside territorial waters is included as part of the overall picture.

Date	Details of incident
3. 9.39	Saro Lerwick (L7252) "BN-?" 240 Squadron. Alighted Dun Laoghaire and later Skerries, Co Dublin. (See p14)
3. 9.39	Short Sunderland (L2158) Marine Aircraft Experimental Estab. Alighted Skerries, Co Dublin. Later this aircraft was on three months loan to BOAC and finally, as (KG-M) with 204 Sqdn, fatally crashed off West Africa. (See p14)
14. 9.39	Short Sunderland (N9023) 228 Squadron. Alighted Ventry Bay, Co Kerry. (See p14)
6. 5.40	Handley Page Hampden 1, Curragh Camp, Co Kildare. Allowed to proceed. (See p15)
20. 8.40	Focke Wolf 200C-1 Condor "F8+KH" I/KG40. Crashed into Faha Ridge, Mount Brandon, Co Kerry. All safe. (See p89)
15. 9.40	Dornier Do18 "K6+FL" III/406. Crashed during attempted take-off from sea off Co Wexford. Crew rescued.
29. 9.40	Hawker Hurricane 1 Z2832 (GZ-M) 79 Sqdn. Out of fuel after combat. Force-landed in Co Wexford. Became "No 93" of Irish Air Corps. (See p39)
11.10.40	Dornier Do172-3 "7T+EK" II Kuestenflieger 606. On Liverpool bombing mission. Shot down by two 611 Squadron aircraft into sea off Co Meath.
22.10.40	Focke Wolf 200C-1 Condor "F8+OK" I/KG40. In sea off Galway. All lost. (See p68)
18.11.40	Blohm und Voss B+V1388-1 "8L+HK" II/Kuestenflieger 906. Missing in sea off Irish coast.
25.11.40	Blohm und Voss BV138A-1 "8L+CK" II/Kuestenflieger 906. Alighted off Blasket Islands, Co Kerry. All safe. (See p70-73)
21.12.40	Miles Master 1 (N8009) 307 Squadron, Isle of Man. Force-landed at Drumgooley, Co Louth. Became instructional airframe, "No 96" of Irish Air Corps.
21.12.40	Bristol Blenheim MkIVF (L9415) 272 Sqdn from Aldergrove. On a reconnaissance mission when it ran out of fuel. The crew of three sergeants baled out successfully though the pilot came down in Lough Swilly and had to spend the night on a rock.
24. 1.41	AW Whitley (T4168), 502 Squadron which was the pre-war NI unit of the RAuxAF. Crashed at Cockavenny, Lifford, Co Donegal. Three lost, two survived when crew baled out over Lough Foyle.(See p48)
24. 1.41	Lockheed Hudson I (P5123) 233 Squadron. Force-landed at Skreen, Co Sligo. Became "No 91" of Irish Air Corps. (See p32)
5. 2.41	Focke Wolfe 200C-3 Condor "F8+AH" I/KG40. Damaged by AA fire from *Major C*. Crashed in dense fog into Cashelfeane Hill, Co Cork. One badly burned crew member survived. (See p69)
23. 2.41	Heinkel He IIIH-5 "1G+LS" VIII/KG27. Shot down into Kenmare River (a bay in Kerry). No survivors. (See caption p88)
3. 3.41	Heinkel He IIIH-5 "1G+HL" I/KG27. Force-landed on Roostoonstown Strand, Co Wexford. (See p83-86)
5. 3.41	Heinkel He IIIH-5 "1G+EK" KG27. Shot down into sea off Stone Head by RAuxAF Sunderland. No survivors. (See p86)
12. 3.41	AW Whitley (P5045) 502 Squadron. Engine trouble, crashed into Galway Bay. Two drowned but two survivors parachuted safely onto land. (See p50)
13. 3.41	Bristol Blenheim (K7068) No5 Bombing and

Gunnery school. Force-landed at Jenkinstown, Co Louth. Pilot slightly injured. (See caption p56)

14. 3.41   Bristol Blenheim 1 (L6720) No5 Ferry Pilot Pool. Force-landed on Clontarf Strand, Dublin city. Polish pilot safe. (See p54-5)

21. 3.41   Consolidated Catalina 240 Squadron. Crashed in flames at Kinlough, Co Leitrim. All killed.

1. 4.41   Heinkel He III H-5 "1G+LH" I/KG27. Force-landed at Bonmahon, Co Waterford. (See p63-6)

8. 4.41   Heinkel He III H-3 "V4+GL" III/KG40. In sea off Co Cork. Two rescued, two lost.

24. 4.41   Fairey Battle TT1 (V1222) "No 8" 4th Bombing & Gunnery School, West Freugh, Scotland. Force-landed on Waterford Racecourse. (See p44)

10. 4.41   Saro Lerwick 1 (L7267) 201 Squadron. Alighted Bundoran, Co Donegal. Stranraer flying-boat brought fuel supply. Both aircraft departed. (See caption p38)

11. 4.41   Vickers Wellington 1 (W5653) 221 Squadron, Limavady. Into Urris Hills, near Fort Dunree, Co Donegal. Depth charge exploded, all six killed. (See top caption p47)

16. 4.41   Consolidated Liberator 4L (755) 103 Squadron Coastal Command.

18. 4.41   Focke Wolfe 200C-3 Condor "F8+GL"III/KG40. In sea off Schull, Co Cork. The aircraft was on fire before ditching. Crew rescued. (See p69 and top caption p68)

18. 4.41   Handley Page Hampden 1 (AD730) "VN-J" 50 Squadron. Returning from German raid. Crashed on Black Hill near Blessington, Co Wicklow. Crew all killed. (See p61-2)

30. 4.41   AW Whitley V (Z6553) 502 Squadron. Crashed at Askill, Co Leitrim. Crew baled out successfully, some into Eire, others into Northern Ireland. (See p50)

6. 5.41   Heinkel He lll H-3 "V4+DK" III/KG27. Shot down into sea near Blackwater Lightship, off Co Wexford. Two rescued, two drowned. (See p86-7)

22. 5.41   Bristol Beaufighter 1C (T3235) 252 Squadron. Force-landed in wheat field near Leopardstown Racecourse, Dublin. (See p21-3)

10. 6.41   Heinkel He lllH-3 "4T+JH" Wekusta 51. Shot down at Churchtown, Co Wexford by Hurricane mentioned below. All killed. (See p39-40)

10. 6.41   Hawker Hurricane IIA (Z2832) "GZ-M" 32 Squadron. Shot down "4T+JH" above before force-landing near Kilmacthomas, Co Waterford. Became "No 94" of Irish Air Corps. (See p39-42)

21. 7.41   Lockheed Hudson I (AM864). Landed on beach at Roskeragh, Co Sligo. Treated as civil aircraft. Took off for Prestwick.

21. 8.41   Hawker Hurricane IIB (Z5070). Ferry flight to NI. Force-landed at Athboy, Co Meath. Became "No 95" Irish Air Corps. (See p42)

26. 8.41   Junkers Ju 88D-2 "4U+HH" I/Aufklarungsgruppe. Shot down and landed at Belgooley, Kinsale, Co Cork. Crew blew up aircraft. (See p73-76)

27. 8.41   Fairey Fulmar (N4072), 804 Merchant Catapult Sqdn, Fleet Air Arm. 'Touch and go' on Tramore Strand, Co Donegal. (See p57-9)

27. 9.41   Lockheed Hudson III (AE577) Ferry flight. Landed Baldonnel then crashed into mountains, Co Louth. All three crew killed. (See p31-2)

2.10.41   Handley Page Hampden I (AD768) "ZM-?". 106 Squadron, Conningsby, Lincs. Crashed into Glendowan Mountains, 10 miles NW of Letterkenny, Co Donegal, after raid on Germany. All parachuted successfully.

11.10.41   Heinkel He lllH-6 "F8+ER" VII/KG40. Shot down and crashed at Rock Mountain, Kiltealy, Co Wexford. All killed.

23.10.41   Bristol Blenheim IV(F) (V5728) "ND-?" 236 Squadron. In sea off Calf Island, Co Cork. Crew of three rescued but one died of injuries. (See p55-56)

25.10.41   Vickers Wellington I "T2506" 103 Squadron. Crew of six baled out successfully. Kilmihill, Co Clare. (See p45-46)

30.11.41   Supermarine Spitfire IIA (B8074) "MD-?" 133 Eagle Squadron, Limavady. At Glenshinny, Co Donegal. Pilot baled out, aircraft crashed deep into bogland. (See p139)

3.12.41   Short Sunderland I (W3988) "ZW-P" 201 Squadron. In sea off Co Clare. Two survivors, nine lost. (See p18-9)

16.12.41   Supermarine Spitfire IIA (P8267). Air delivery flight from Isle of Man to Limavady. Force-landed on Maghermore Strand, near Clogher, Co Donegal. (See p38)

21.12.41   Grumman Martlet II (AM975) 881 Squadron, Fleet Air Arm. Cloghfin, near Carrigans, Co Donegal after engine trouble during flight off HMS *Illustrious* to Rugby. (See p60)

21.12.41   Vickers Wellington I (Z1145). Landed at Gormanston aerodrome, refuelled and allowed to leave.

26.12.41   Junkers Ju 88A-5 "DE+DS" Wekusta 2. Force-landed in bog near Waterville, Co Kerry. All safe. (See p94-6)

28. 1.42   Hawker Hurricane I 59 OTU. American pilot landed Dublin Airport. (See p81)

5. 2.42   Short Sunderland (W3977) "ZM-Q" 201 Sqdn. Into sea nine miles off Dunmore Head, Co Donegal. Twelve killed. (See caption p19)

24. 2.42   Lockheed Hudson III (AM834) No1 OTU. In sea off Cahore Point, Co Wexford.

3. 3.42   Junkers Ju 88A-4 "CN+DU" Wekusta 2. Into Mount Gabriel, Co Cork. (See p93)

5. 3.42   Lockheed Hudson (FH263). Near Blacksod Bay en route Gander to Prestwick. Only two gallons fuel left.

6. 3.42   Catalina Mk I (VA-721) from British Ferry Command came down in blizzard conditions off Malin Head. Boatman contacted a Royal Navy destroyer and directed it to the Catalina whose crew were rescued.

13. 3.42   Bristol Blenheim l (Z6021) "B6" of No5 Bombing and Gunnery School. Force-landed Crossmolina, Co Mayo. Three injured. (See p53-4)

16. 3.42   Consolidated Liberator (AL-755) 103 Squadron. On flight from Egypt to UK, crashed into Dundalk Mountains, Co Louth.

21. 3.42   Heinkel He IIIH-6 "F8+ET" IX/KG40. Shot down in sea. Crew in dinghy reached Bull Rock Lighthouse. (See p87-8)

15. 4.42   Lockheed Hudson. Ferry flight. Into sea off Binghamstown, Co Mayo. Three crew rescued.

19. 4.42   Bristol Blenheim. (N3533) 143 Squadron, Aldergrove. On a training exercise with the British Army. Lost bearings and low on fuel, force-landed at Glebe, near Buncrana, Co Donegal.

24. 4.42   Curtiss Tomahawk I (AH920) RAF trainer. Force-landed Kilcarra, Co Wicklow. Aircraft repaired at Baldonnel; pilot briefly interned. (See p117-8)

4. 5.42   Lockheed Hudson (FH376). Ferry flight, Canada to Scotland. Landed at Ballyliffen, Co Donegal and allowed to proceed.

25. 5.42   Lockheed Hudson (FH233). On ferry flight. Into sea at western end of Sligo Bay. Crew of three drowned. Co Mayo.

16. 6. 42   Lockheed Hudson (US bureau code 426613). RAF ferry flight. Landed on Hill Strand, Dunfanaghy, Co Donegal. Refuelled and allowed to depart.

27. 6.42   Junkers Ju 88D-5 "4T+MH" Wekusta 51. In sea 15 miles south of Hook Head. (See p65)

4. 7.42   North American B-25 Mitchell. Ferry flight. Force-landed, Riversdale, Co Roscommon. (See p141)

6. 7.42   Douglas C-47 Skytrain. Ferry flight to Prestwick. Landed Rineanna.

17. 7.42   Bristol Beaufort (N1063) 5 OTU, Long Kesh. Crashed on Ballyness Strand, Co Donegal. Crew set fire to aircraft which burnt out. One crew member native of Donegal.

25. 7.42   Westland Lysander (2745) Valley, Wales. Force-landed on Ballyliffen Strand, Co Donegal. Refuelled and allowed to depart.

13. 8.42   Consolidated B24 Liberator (LV341) 120 Sqdn, Ballykelly. Crashed into sea 37 miles off Tory Island, Donegal. All crew killed.

23. 8.42   Spitfire 315 (Polish) "Deblinski" Squadron, RAF. Pilot wounded in combat with following aircraft and crash-landed in Co Meath. Died from wounds and crash injuries. (See p77)

23. 8.42   Junkers Ju 88D-1 "4U+KH". Shot down by Spitfires of 315 Squadron. Crash-landed at Tougare, Tramore, Co Waterford. (See p77-80)

12. 9.42   Vickers Wellington. Into Lough Foyle on Border centre line. Crew rescued by British armed trawler.

20. 9.42   Spitfire V No140 Met Flight. Figgular, Co Monaghan.

26. 9.42   Wellington (R1021). Crash-landed at Charleville, Co Cork. All safe. (See caption p48)

30. 9.42   Lockheed Ventura (AJ460). Ferry flight which force-landed at Roonagh Point, Co Mayo. (See p16-8)

25.10.42   Lockheed Ventura crashed at Cloonooragh, Co Mayo.

25.10.42   Douglas A-20 Boston (BZ200) Overturned on crash-landing at Crossmolina, Co Mayo. Pilot drowned. (See p120-1)

29.10.42   Handley Page Hampden 5 OTU, Long Kesh. On navigation training flight, crashed close to Ballybofey, Co Donegal. All crew safe. (See p62)

30.10.42   Spitfire Vb (BM553) "EB-M" 41 Squadron. Force-landed at Oulart, Co Wexford. Polish pilot slightly injured.

17.11.42   Consolidated Catalina (SB273). Ferry flight to Largs, Scotland. Alighted on Lough Gill, Co Sligo. Allowed to proceed. (See p25)

19.11.42   Consolidated Catalina Mk IB (FP202) No 302 Ferry Training Unit. Alighted at Foynes. (See p24)

26.11.42   Spitfire Vc (AD116) "SD-?" 501 Squadron, Ballyhalbert. On training flight. Stranorlar, Co Donegal.

5.12.42   Boeing B-17 Fortress 'The Devil Himself'. Mullaghmore, Co Sligo. Lengthy repairs, then proceeded to UK base. (See p126-7)

8.12.42   Miles Mentor. Phoenix Park, Dublin city. US pilot returned to NI by US Embassy chauffeur.

17.12.42   Lockheed P-38 Lightning. Crashed at Corrintra, Castleblaney, Co Monaghan.

18.12.43   Boeing B-17 Fortress USAAF (4337895). Bad weather caused aircraft to land at Rineanna.

23.12.42   Lockheed P-38 Lightning. Force-landed Ballyvaughan Strand, Galway Bay, Co Clare. (See p115)

2. 1.43   Vickers Wellington (HX467) 7 OTU, Limavady. Crashed into Lough Foyle. All six killed.

9. 1.43   Lockheed P-38 Lightning (4212802). Floated off torpedoed cargo vessel. Eventually into Lough Foyle, Co Donegal. (See p117)

15. 1.43   Boeing B-17E Fortress (419045) 'Stinky'. VIP flight from North Africa. Force-landed at Athenry, Co Galway. (See p111-115)

6. 2.43   Aircobra (424518) "ZL-5" 427 USAAF Force-landed at Campile, Co Wexford. (See p120)

17. 2.43   Vickers Wellington III (Z1676) "ZL-S" 427 (Lion) Squadron. Down at Carricklong, Co Waterford. Aircraft destroyed by crew. (See p46-7)

27. 2.43   Vickers Wellington VXIII (HX737) 7 OTU. Into bog near Falcarragh, Co Donegal. All six killed.

1. 3.43   Vickers Wellington III. Near Waterford city.

17. 3.43   Bristol Beaufighter. St Mogue's Island, Co Cavan. Crew of two parachuted safely into NI.

18. 3.43   Consolidated B-24 Liberator (FK222) 86 Sqdn, Aldergrove. On Tallan Strand, near Fort Lenan, Co Donegal. All crew safe. (See captions p34)

21. 3.43   Boeing B-17F Fortress (205220). On a delivery flight via North Africa. Landed Rineanna. (See p129)

7. 4.43   Boeing B-17 Fortress (723090) 'T'aint a Bird'. Lost and out of fuel. Conventional landing in marshy land. Took off again on specially laid metal runway. Clonakilty, Co Cork (See p108-11)

17. 4.43   Boeing B-17F Fortress (423143). Ferry flight from Newfoundland to Littleport aerodrome, England, via North Africa. Landed Dublin Airport. (See p81)

10. 5.43   Boeing B-17G 20-VE Fortress. Ferry flight. Portnablagh Strand. No casualties. (See p127-8).

25. 5.43   Short Sunderland (DD846) "2-D" 422 Squadron RCAF. Glanced off Clare Island. Into sea Clew Bay, Co Mayo. All crew killed. (See p20)

4. 6.43   Martin B26 Marauder 'Ridge Runner'. Force-landed on Termonfeckin Strand, Co Louth. (See p119)

29. 6.43   Avro Anson I (LT985) 9 (Observer) Advanced Flying Unit. Landed Dublin Airport. (See p82)

10. 7.43   RAF Catalina into sea off Innishowen Head. Crew picked up by RN destroyer.

10. 7.43   Consolidated B-24D Liberator (4240784) 'Travellin' Trollop'. Ferry flight with combat crew. Force-landed Lahinch Strand, Co Clare. (See p101-3)

16. 7.43 DH Albatross G-AFDK 'Fortuna' of BOAC. Broke up coming in to land at Rineanna. No casualties. (See p130-1)

17. 7.43 Vickers Wellington III (X3563) "ZL-T". Polish crew on anti U-boat mission over Bay of Biscay. On return fog encountered and fuel low. Crew baled out successfully over Ballinlough, Co Roscommon. (See p47)

23. 7.43 Junkers Ju 88D-1 "D7+DK" Wekusta 2. Crashed in fog on Bellinacarriga Hill, Dursey, Co Cork. All killed. (See p94)

28. 7.43 Short Sunderland III BOAC "G-AGES". Mount Brandon, Co Kerry. Ten killed. (See p91-2).

16. 8.43 Bristol Beaufort. Force-landed at Ballyness Strand, near Falcarragh, Co Donegal. Unusually this aircraft had a Fleet Air Arm crew.

22. 8.43 Short Sunderland III (DD848) 201 Squadron. Into Mount Brandon, Co Kerry. Eight killed. (See p92)

27. 8.43 Consolidated B-24 Liberator RAF. Total wreck, main elements destroyed by the Ordnance Corps. Castletownbere, Co Cork.

16. 9.43 Lockheed Hudson V (AM885) 301 Flying Training Unit. Crashed Ballyculhane, Co Wexford. All killed.

12.10.43 Avro Anson I (LV139) "ED-4" 12 Air Gunnery School, Bishopscourt. NI Landed Dublin Airport. (See p82)

5.11.43 Two C-47's (224074) and (224098). Both aircraft lost bearings and landed at Rineanna. (See p130)

7.11.43 Handley Page Halifax (EB-134) 1663 Heavy Conversion Unit. Crash-landed Tuam, Co Galway. All killed. (See p51)

13.11.43 Short Sunderland (DD863) 423 Squadron. In sea off Donegal coast. Eleven killed.

9.12.43 Boeing B-17 Fortress USAAF. Out of fuel on ferry flight. Force-landed on beach near Belmullet, Co Mayo. Salvage attempt by US engineers caused further damage, and aircraft was dismantled by an Air Corps team.

11.12.43 Boeing B-17G Fortress (4231320). Ferry flight. Truskmore Mountains, Co Sligo. (See p123)

13.12.43 Focke Wolf 200C-3 Condor "F8+MR" VIII/KG40. Returning from sortie to Iceland.

Force-landed at Dromineer, Co Tipperary. All safe. (See p70)

14.12.43 HP Harrow (K7005), 271 Squadron en route Eglington, Derry to Doncaster. In dense fog crashed 5 miles NW Moville, Co Donegal. Only one survivor of five crew members.

17.12.43 Grumman Martlet FAA. St Johnston, Co Donegal.

17.12.43 Douglas C-47 Skytrain (4330719) 437 Transport Carrier Group. Into Carrantohill, McGillicuddy Reeks, Co Kerry. Five killed. (See p104-5)

18.12.43 Boeing B-17G Fortress (237895). Landed Rineanna. (See p129)

20.12.43 Vickers Wellington XVI (HF208) 304 (Polish) Selesion Squadron. Mount Brandon, Co Kerry. (See p92)

24.12.43 Boeing B-17 Fortress (427443). Urris Hills, Co Donegal. All safe.

28.12.43 Miles Martinet TT (HP371) 131 OTU. Crashed at Scotstown, Co Monaghan. Pilot killed.

23. 1.44 Boeing B-17G Fortress (4331507). Crewless – all crew had bailed out over NI Aircraft landed at Johnstown, Co Kilkenny. (See p123)

23. 1.44 Handley Page Halifax (LK714) No 517 Met flight. From Tiree, Scotland. Crashed into cliffs above Tullan Strand, Bundoran, Co Donegal. All killed. (See p51 and 53)

24. 1.44 Boeing B-17 (--7443). Erris Head, Co Mayo.

31. 1.44 Handley Page Halifax. Force-landed at Doolough, Co Mayo.

31. 1.44 Short Sunderland III (DW110) 228 Squadron, Pembroke Dock, Wales. Re-routed to NI and crashed into Blue Stack Mountains, Co Donegal. Three survivors. (See p20-1)

4. 2.44 Consolidated B24 Liberator (4252404). Landed Rineanna. Departed 10th February. (See p130)

8. 2.44 Boeing B-17G (4231971) USAAF. En route Goose Bay to UK. Landed Rineanna. (See p130)

11. 2.44 North American B25 Mitchell (FW235) USAAF ferry flight Gander to Prestwick. Out of fuel. Doolough, Co Mayo.

20. 2.44 Boeing B-17 Fortress (427533). Gander to UK. Force-landed on north shore of Donegal Bay. Aircraft total wreck, all ten crew injured.

22. 2.44 Martin B26B Marauder (4295944). En route Marrakesch to NI. Landed Rineanna. (See p130)

22. 2.44 Douglas DC-3 Skytrooper. Leopardstown Racecourse, Dublin. Took off after runway prepared by Army Engineers. (See p103-4)

23. 2.44 Consolidated B24J Liberator (2109825). On route from Marrakesch to N Ireland. Landed Rineanna. (See p130)

27. 2.44 Consolidated PB4Y (63939). Air Fleet 7. Hit Skellig Rock, off Co Kerry. All killed. (See p106)

5. 3.44 Douglas C-47 (4223395). Landed Rineanna.

9. 3.44 Consolidated Catalina MkIVB (JX330). Foynes. (See p24)

12. 3.44 Supermarine Seafire (LR-841). Landed Gormanston aerodrome. (See p60)

12. 3.44 Armstrong Whitworth Whitley. Landed Dublin Airport. (See p82)

16. 3.44 Consolidated Liberator. Force-landed at Skibbereen, Co Cork. One crew member killed.

17. 3.44 Lockheed P-38 Lightning. Force-landed at Gormanston aerodrome.

17. 3.44 Lockheed P-38 Lightning. Force-landed near Monaghan town.

21. 4.44 Halifax (LL-145) 517 (Met) Squadron. Wheels down landing at Skibbereen, Co Cork. RAF Maintenance Unit allowed to dismantle which became a frequent concessions from about this date. (See p53 and captions p52)

22. 4.44 Avro Anson "B-3" No 7 Air Observer School, Bishopscourt. Night landing at Baldonnel.

25. 4.44 Avro Anson I (DJ639) No 10 Radio School. Into sea of Tuskar Rock, five miles from Wexford coast.

5. 5.44 Consolidated B-24. Ferry flight. Crash-landed at Foxford, Co Mayo.

5. 5.44 Two Piper L-4 Grasshoppers. Landed at Stackallen, Co Meath, checked position, allowed to depart. (See p121)

5. 5.44 Lockheed P38 Lightning. Force-landed at Gormanston aerodrome. Aircraft badly damaged but pilot uninjured. (See p117)

29. 5.44 Boeing B-17 Fortress (423279) USAAF 'Badger Beauty'. On Met flight, ditched off Dingle, Co Kerry. All safe. (See p125-6)

9. 6.44 Fairey Swordfish, Fleet Air Arm. Down in sea. Crew took to dinghies which were blown out to sea but all three were rescued by LOP personnel. Off Achill Head, Co Mayo.

19. 6.44 Consolidated Liberator VIII (FL989) 59 Sqdn, Ballykelly. Crashed into Glengad Head, Co Donegal. No survivors. (See p38)

19. 6.44 Consolidated Liberator VIII (FL990) 59 Sqdn, Ballykelly. Crashed into Shrove Hill, Co Donegal. No survivors. (See p38)

19. 6.44 Consolidated B-24 Liberator. USAAF Ferry flight. Crashed near Ballyshannon, Co Donegal. Two crew members killed. (See p38)

23. 6.44 Fairey Swordfish. Fleet Air Arm. Landed near Elly Bay, Co Donegal.

13. 7.44 Consolidated B-24 Liberator (BZ910). In sea close to Innistrahull, Co Donegal.

20. 7.44 Lockheed P38 Lightning F-5a photographic reconnaissance version 'Shoo-Shoo-Baby'. Clonsast Bog, Co Offaly. (See p115-6)

12. 8.44 Short Sunderland III (NJ175) "DT-G" (The Flying Yachtsmen) 422 Squadron RCAF. Engine failure near Belleek, Co Donegal. Three killed.

1. 9.44 Fairey Swordfish. Landed at Gormanston aerodrome after becoming lost on communications flight from Isle of Man. (See p59)

4. 9.44 Fairey Swordfish. Force-landed at Carrowcastle, Co Sligo. (See p59-60)

5. 9.44 Consolidated Catalina (JX422). Foynes. (See p24)

6. 9.44 Short Sunderland 423 Squadron RCAF. Into Donegal Bay. Nine killed.

14. 9.44 Consolidated PB4Y (38799) US Navy. Ferry flight. Into sea off Clifden, Co Galway. Three survivors. (See p106-7)

14. 9.44 North American Mustang crashed at Kippans, Co Louth. Aircraft appears to have power-dived into a field, creating a wide 6ft deep crater.

4.10.44 Vickers Wellington "HF450". Into Galway Bay.

9.12.44 Boeing B-17G Fortress (20-VE) 'Bar Fly'. Ferry flight to 303 Bomber Group. Crashed at Ballintrillick, Co Donegal. (See p126)

17.12.44 Martinet "GB-HP370" Fleet Air Arm. Target towing aircraft. Force-landed in Co Sligo.

20.12.44 Consolidated Catalina IV (JX208) "TQ-F" 202 Squadron. Crashed at Castlegregory, Co Kerry. All lost. (See p25)

9. 2.45 Handley Page Halifax (MZ980) 298 Squadron. On navigational exercise. In sea off Mullaghmore, Co Sligo. (See p53)

10. 2.45 Consolidated Liberator (KK295). 45 Group ferry flight, force-landed on strand at Portsalon, Co Donegal. No fatalities.

15. 2. 45 Martin B26G Marauder (4468079). Force-landed at Gorey, Co Wexford. (See p118)

10. 3.45 Avro Anson I (MK241). Force-landed in bogland and could not be salvaged. Ballynaderreen, Co Roscommon. (See p81-2)

14. 3.45 Short Sunderland (ML743) 201 Squadron. Crashed near Killybegs, Co Donegal. Twelve killed. (See p19-20)

14. 3.45 Hawker Hurricane IIC (PZ774) 1402 Met flight, Aldergrove. Force-landed five miles from Moville, Co Donegal.

5. 4.45 Avro Anson I 7 (Observer) AFU, Bishopscourt. Landed Dublin Airport. (See p82)

7. 4.45 Supermarine Spitfire XIX (EN409) "BE-?" No 8 OTU (Photo-Recce). All over blue finish. Force-landed Spanish Point, Miltown Malbay, Co Clare. (See p99-100)

5. 5.45 Junkers Ju 88G-6c "D5+GH" 1/NJG3. Landed at Gormanston aerodrome. Became "DK888" when taken over by RAF. (See p96-9)

21. 5.45 Short Stirling glider tug. Force-landed at Gormanston aerodrome. (See caption p81)

24. 6.45 Grumman Hellcat (JV176). Force-landed on Greencastle golf course, Co Donegal. (See captions p58-9)

27. 6.45 Fairey Barracuda II (MD897) "R5D" FAA. Rosslare, Co Wexford. (See p60-1)

28. 6.45 De Havilland Mosquito FB26. Atlantic Ferry flight. Hit stone cairn at Clifden, Co Galway. One killed, one survivor.

29. 7.45 Lockheed Embassy. Political visit. Force-landed at Ballina, Co Mayo.

3. 1.46 Grumman Martlet. Training flight from NI. Lord Mountbatten's estate at Mullaghmore, Co Sligo.

29.11.46 Avro Anson. No 7 Air Navigation School, Bishopscourt. Into Limerick Docks. Sank after salvage of equipment. (See p82)

Author's Note

Donal MacCarron would be glad to hear from readers in regard to additions and amendments to the above list. His address is 4 The Chyne, Gerrards Cross, Bucks. SL9 8HZ, England. Tel/Fax: 01753 883812. Email: donmacs@onetel.com

Index

Italics indicate a picture of, or relating to, the subject.

People

Places

Landfall Ireland

Planes

Ships